Vipin Kumar (Ed.)

Hybrid Composite Materials and Manufacturing

Also of Interest

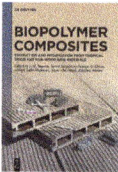

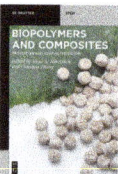

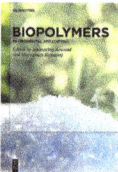

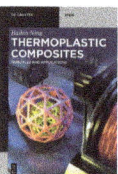

Hybrid Composite Materials and Manufacturing

Fibers, Nano-Fillers and Integrated Additive Processes

Edited by
Vipin Kumar

DE GRUYTER

Editor
Dr. Vipin Kumar
Oak Ridge National Laboratory
1 Bethel Valley Road
P.O. Box 2008
Oak Ridge, TN 37831
United States of America
kumarvi@ornl.gov

ISBN 978-3-11-101934-5
e-ISBN (PDF) 978-3-11-101954-3
e-ISBN (EPUB) 978-3-11-102013-6

Library of Congress Control Number: 2024943796

Bibliographic information published by the Deutsche Nationalbibliothek
The Deutsche Nationalbibliothek lists this publication in the Deutsche Nationalbibliografie;
detailed bibliographic data are available on the internet at http://dnb.dnb.de.

© 2025 Walter de Gruyter GmbH, Berlin/Boston
Cover image: Hybrid additive manufacturing-compression molding (AM-CM) technology, Department of
Energy's Manufacturing Demonstration Facility at Oak Ridge National Laboratory
Typesetting: Integra Software Services Pvt. Ltd.

www.degruyter.com
Questions about General Product Safety Regulation:
productsafety@degruyterbrill.com

Preface

The evolution of composite materials has revolutionized multiple industries, which are constantly looking for lightweight structural materials. The composite evolution is driven by the innovative synergy of combining distinct materials and manufacturing processes to achieve superior properties. As Mark Twain once described, "the bonus that is achieved when things work together harmoniously" epitomizes the essence of composite technology. This fundamental principle, inspired by nature and exemplified in structures like human bone, highlights the profound benefits of material integration. However, nature provided few clues for optimized composite manufacturing processes that we must innovate together. Composites are as much an art (manufacturing processes) as science (materials); without a proper understanding of both, composites will never reach their true potential. This book has tried to explain both aspects.

With escalating technological demands, the pursuit of enhanced composite materials remains relentless. Researchers worldwide are dedicated to discovering new properties and refining manufacturing techniques to meet these challenges. Hybridization emerges as a leading strategy, merging the advantages of diverse materials and processes to transcend their individual limitations. Hybrid composites, which integrate multiple matrix materials or reinforcing fillers and employ hybrid manufacturing processes, are at the forefront of this innovative approach.

This book delves into the latest research and industrial progress in hybrid composite materials and manufacturing techniques. It covers a comprehensive range of topics, including hybrid matrix materials, hybrid fiber composites, advanced manufacturing processes, and nanocomposites. The unique aspect of this book is the detailed explanation of recent trends in integrated processes, where traditional manufacturing processes are married with recently developed advanced manufacturing processes. Our goal is to provide readers with a thorough understanding of the current state and prospects of hybrid composites, ensuring they are well-informed and equipped with the latest knowledge.

The authors explored the intricate interplay of different polymers, the precise tuning of processing parameters, and pioneering synthesis and blending methods. From developing hybrid matrix materials and interpenetrating polymer networks to creating hybrid fiber composites and incorporating nanoscale materials, this book offers in-depth insights essential for advancing composite technology.

By presenting these advancements, I aim to inspire ongoing innovation and practical applications of hybrid composites across various sectors. This compilation is a valuable resource for academic researchers and industry professionals engaged in the dynamic field of polymer composites.

I extend my heartfelt gratitude to the researchers and students whose contributions have made this book possible. Their unwavering commitment to excellence continues to drive the field forward, paving the way for future breakthroughs in hybrid

https://doi.org/10.1515/9783111019543-202

composite materials and manufacturing. Last but not least, I want to express my most sincere gratitude to my wife, my parents, and my newborn boy, Rei Kumar. I hope one day Rei will read this book and be inspired by the exciting world of composites.

Vipin Kumar, Ph.D.
Knoxville, TN, USA
July 2024

Contents

Contributors

Subhabrata Saha
Manufacturing Science Division,
Oak Ridge National Laboratory,
Knoxville, TN
USA

Komal Chawla
Manufacturing Science Division,
Oak Ridge National Laboratory,
Knoxville, TN
USA

Sanjita Wasti
Tickle College of Engineering
University of Tennessee Knoxville
Knoxville, TN
USA

Katie Copenhaver
Manufacturing Science Division
Oak Ridge National Laboratory
Oak Ridge, TN
USA

Xianhui Zhao
Environmental Sciences Division
Oak Ridge National Laboratory
Oak Ridge, TN
USA

Umesh Marathe
Manufacturing Science Division
Oak Ridge National Laboratory
Oak Ridge, TN
USA

Abdallah Ragab Barakat
Tickle College of Engineering
University of Tennessee Knoxville
Knoxville, TN
USA

Surbhi Kore
Intel
Hillsboro, OR
USA

Soydan Ozcan
Manufacturing Science Division
Oak Ridge National Laboratory
Oak Ridge, TN
USA

Uday Vaidya
Tickle College of Engineering
University of Tennessee Knoxville
Knoxville, TN
USA
and
Manufacturing Science Division
Oak Ridge National Laboratory
Oak Ridge, TN
USA
and
Institute for Advanced Composites
Manufacturing Innovation (IACMI)
Knoxville, TN
USA

Sangita Tripathy
Academy of Scientific and Innovative Research
(AcSIR)
Ghaziabad 201002
India
and
CSIR-National Physical Laboratory
New Delhi 110012
India

S. R. Dhakate
Academy of Scientific and Innovative Research
(AcSIR)
Ghaziabad 201002
India
and
CSIR-National Physical Laboratory
New Delhi 110012
India

Bhanu Pratap Singh
Academy of Scientific and Innovative Research
(AcSIR)
Ghaziabad 201002
India

https://doi.org/10.1515/9783111019543-204

and
CSIR-National Physical Laboratory
New Delhi 110012
India

Umesh Marathe
Manufacturing Science Division
Oak Ridge National Laboratory
Oak Ridge, TN
USA

Georges Chahine
Tickle College of Engineering
University of Tennessee Knoxville
Knoxville, TN
USA

Chase McCullar
Tickle College of Engineering
University of Tennessee Knoxville
Knoxville, TN
USA

Pritesh Yeole
Tickle College of Engineering
University of Tennessee Knoxville
Knoxville, TN
USA

Sana Elyas
Manufacturing Science Division
Oak Ridge National Laboratory
Oak Ridge, TN
USA

Kazi Md Masum Billah
Assistant Professor of Mechanical Engineering
University of Houston–Clear Lake
Houston, TX
USA
email: billah@uhcl.edus

Steve Bullock
Manufacturing Science Division
Oak Ridge National Laboratory
Oak Ridge, TN
USA

Tony Beard
Manufacturing Science Division
Oak Ridge National Laboratory
Oak Ridge, TN
USA

David Nuttall
Manufacturing Science Division
Oak Ridge National Laboratory
Oak Ridge, TN
USA

Akash Phadatare
Tickle College of Engineering
University of Tennessee Knoxville
Knoxville, TN
USA

Ahmed Arabi Hassen
Manufacturing Science Division
Oak Ridge National Laboratory
Knoxville
TN 37932
USA

Segun Isaac Talabi
Manufacturing Science Division
Oak Ridge National Laboratory
Knoxville
TN 37932
USA

Vipin Kumar
Manufacturing Science Division
Oak Ridge National Laboratory
Knoxville
TN 37932
USA

Subhabrata Saha, Komal Chawla, Vipin Kumar

1 Introduction

1.1 Definition

Synergism, *"the bonus that is achieved when things work together harmoniously"* (by Mark Twain), is the primary working principle of composites. The concept of composites came from natural sources, and the most common example is the human bone. Composites typically consist of at least two different materials with a distinct phase boundary. In the field of polymers, the idea of developing composites is to replace metals that are difficult to deal with in terms of materials handling and processing. Polymers usually form the matrix phase, which provides structural integrity, while the reinforcing material, either in the form of a particulate, fiber, or fabric, contributes to the strength. Nowadays, one can see the use of composites almost every corner, from households to construction, automotive, marine, wind, aerospace, and biomedical fields.

Nonetheless, technological advancement demands further upgrading of composite materials to introduce multifunctional properties, robustness, higher strength, intricate shapes, and complex architecture in conjunction with fast manufacturing and low production costs. Worldwide, researchers from academia and industries are continuously working on improving properties and manufacturing methods to extend the application area further. Hybridization is one of the promising solutions to combine the advantages of parent technologies and the simultaneous attenuation of their undesirable drawbacks. In a broader sense, hybrid composites include hybrid materials and manufacturing technologies. Hybrid composite materials refer to those that contain more than one combination of matrix materials and reinforcements. When two or more conventional manufacturing methods are integrated (e.g., injection-compression molding, additive manufacturing-compression molding, etc.), the process is called hybrid manufacturing. The book combines recent research outcomes and the latest industrial technologies for hybrid materials and manufacturing techniques in polymer composites.

Notice: This manuscript has been authored by UT-Battelle, LLC, under contract DE-AC05-00OR22725 with the US Department of Energy (DOE). The US government retains and the publisher, by accepting the article for publication, acknowledges that the US government retains a nonexclusive, paid-up, irrevocable, worldwide license to publish or reproduce the published form of this manuscript, or allow others to do so, for US government purposes. DOE will provide public access to these results of federally sponsored research in accordance with the DOE Public Access Plan (https://www.energy.gov/doe-public-access-plan).

Subhabrata Saha, Komal Chawla, Vipin Kumar, Manufacturing Science Division, Oak Ridge National Laboratory, Knoxville, TN USA

https://doi.org/10.1515/9783111019543-001

The term "hybrid matrix composites" is used when combining more than one matrix. The matrix material includes thermoplastics, thermosets, and bio-derived polymers. Different morphologies can be formed when multiple matrix materials are blended, such as co-continuous, droplet-matrix, lamellar, etc., ultimately defining the final properties. Factors like solubility and interaction parameters, volume or weight fractions, processing parameters (temperature, pressure, and shear force), etc., govern the morphology development. Similarly, hybridization in the reinforcing phase can be done by combining different fibers, fibers/particulate fillers, fibers/fabrics, etc. The use of multiple fibers in fiber-reinforced polymer composite (FRP) has gained significant research interest in recent years, and these materials are known as hybrid fiber composites. There are two basic classifications of hybrid composites. Class I is the category where Van der Waals forces, H-bonded interaction, or electrostatic interaction bond different components in the hybrid composites. In Class II hybrid materials, chemical bonding forms the hybrid structure [1]. According to Nanko et al., hybrid materials can be divided into three categories: structurally hybridized materials, materials hybridized by chemical bonds, and functionally hybridized materials [2].

The International Academy for Production Engineering (CIRP) proposed that a process be considered hybrid manufacturing when two or more established manufacturing processes are combined into a new setup where the advantages of each discrete process can be exploited synergistically [3]. Another narrow definition of hybrid processes is the simultaneous action of different processing principles on the same processing zone [3]. Combining multiple processes offers the flexibility to manufacture complex parts and maintain design accuracy in a relatively short production time.

1.2 Advantages of hybrid composite

Hybrid composites belong to the class of multifunctional materials capable of offering more than one characteristic benefit. In structural applications, hybrid structures like metal-polymer composites, metal matrix composites, over-molded parts, etc., provide several advantages in stiffness-to-weight ratio, load transfer, stress distribution, thermal stresses, toughness, etc. Hybrid metal-polymer laminates containing FRP inserted with metal parts (e.g., aluminum (Al), titanium (Ti), etc.) are being widely used in passenger aviation industry, which collectively provide higher specific strength, fatigue performance, impact properties, and vibration resistance [4]. Hybrid fiber composites are another famous name in the composite society. A typical example is carbon fiber/aramid fiber composites, where the CF governs the strength and stiffness while aramid fiber provides toughness [5]. Hybrid composites can also offer a cost-cutting solution by combining less expensive materials without or marginally sacrificing the properties. In recent years, the trending nanohybrid composites containing nanostructured materials exhibit a drastic improvement in functional properties, including electrical and thermal conductivities, energy absorption, surface properties, antifouling properties, optical and barrier properties, etc. These nanohybrid composites are widely studied in high-end so-phisticated applications like sensors, fuel cell membranes, solar cells, drug delivery, ca-pacitors, catalytic applications, optoelectronics, wastewater management, etc. Hybrid composites also contribute to sustainability issues by combining bio-derived and syn-thetic materials. The combination of natural fibers/synthetic fibers to form biohybrid composites is getting enormous attention in minimizing the usage of petroleum feed-stocks. For example, bio-based sisal fiber/glass fiber/polypropylene (PP) hybrid compo-sites exhibit an excellent combination of strength, cost, and water-resistant properties [6, 7]. Hemp fiber (natural plant fiber)/glass fiber composites significantly reduce costs and improve mechanical and physical properties [7]. Incorporating bamboo-cellulosic fiber in the PP/polylactic acid (PLA) blend improves the fracture strength [8]. Thanks to these benefits, hybrid composites can be promising across various industries for a wide variety of products.

1.3 Hybrid matrix materials

The most common way to develop hybrid matrix material is by mixing two distin-guishable polymeric molecules. Mixing methods include mechanical blending, block-copolymerization, graft copolymerization, and semi-interpenetrating networks. Two or more different polymers can be melted or mixed by mechanical shearing or agita-tion. The polymers involved in melt mixing should have similar melting/softening temperatures, or at least the melting/softening temperature of one of the polymers should not lie in the degradation temperature region of the other polymer. For solu-tion blending, the polymers must have close solubility parameter values; otherwise, there will be phase separation in the solution mixture. In this case, viscosity and mo-lecular weight play a significant role in achieving uniform mixing. In general, no chemical bonds are formed between the constituent polymers in the case of polymer blends. However, mechanical shearing also generates free radicals through molecular degradation. The mechanochemically generated radicals subsequently react to form chemical grafting between the components. Blends can be considered miscible or im-miscible depending on the sign of the free energy of mixing. Different classes of poly-mers often need to be blended to achieve diverse properties, forming immiscible blends. Compatibility is required for immiscible blends to reduce interfacial tension, stabilize morphology, and enhance adhesion between solid phases. The following strategies are usually adopted for compatibilization: (i) the addition of a small quan-tity of a third component, which typically contains two different segments soluble in two different phases; (ii) the addition of core-shell polymers that behave like a compa-tibilizer as well as an impact modifier; (iii) reactive blending, e.g., dynamic curing, where cross-linking occurs during mixing [9].

Block copolymers comprise a linear arrangement of blocks in which the constitutional repeat unit of one block differs from the adjacent blocks. Each block type has distinct characteristics that, in combination, give rise to a judicious set of properties. Block copolymers can be synthesized by chain growth (radical) polymerization as well as step polymerization (condensation) [10]. The sequential addition of different monomers or chain extensions from dual initiators is the chain polymerization method to synthesize block copolymers. Living free radical polymerization is a convenient way to sequentially add a monomer as the propagating radicals do not undergo a chain termination reaction, unlike conventional free radical polymerization, where the polymerization proceeds via chain-breaking and chain-transfer mechanisms. Atom transfer radical polymerization (ATRP), nitroxide-mediated polymerization (NMP), and reversible addition-fragmentation chain-transfer polymerization (RAFT) are the various types of living radical polymerization techniques frequently used for synthesizing block copolymers. The successive addition of monomers can be done in two ways: one-pot synthesis and isolated macroinitiator methods [11]. In one-pot synthesis, the second monomer is added after the 90% conversion of the first monomer. The drawback of the one-step method is the contamination of the first monomer unit in the second block. In isolated macroinitiator methods, the first block obtained by homopolymerization is isolated after 50% conversion, commonly by precipitation, and later, the homopolymer block acts as a macroinitiator in forming the second block. Another approach is the polycondensation method, which synthesizes a telechelic macroinitiator, which later forms a triblock copolymer. For example, condensation-polymerized hydroxyl-terminated blocks are reacted with the chain transfer agents (CTA) to form macro-CTA, which further undergoes ring-opening polymerization or living radical polymerization to form ABA triblock copolymer [11]. Typical examples of block copolymers are polystyrene-b-polyacrylonitrile (SAN), polystyrene-b-polymethyl methacrylate, polyvinyl pyrrolidone-b-polystyrene, etc.

A graft copolymer contains a long homopolymer sequence as a backbone with branches of a long sequence of another monomer [10]. There are three basic methods of synthesizing graft copolymers: (1) grafting through two functional groups of two different macromolecular units. In this method, one macromolecular unit contains side chain functional groups, while the other macromolecular unit has end-terminal functional groups; (2) grafting through the side chain functional group of a polymer, which acts as a macroinitiator and initiates the polymerization of the second monomer; (3) grafting through copolymerization using a vinyl macromonomer. Depending on the grafting density and shape, the graft copolymers are designated as comb polymers, brush polymers, and star-shaped polymers.

An interpenetrating polymer network (IPN) is a combination of two or more polymers interlaced at a molecular level without forming any covalent bonds between each other; however, they cannot be separated unless the chemical bonds are ruptured [12]. The IPN structure is formed when at least one of the polymers is synthesized or cross-linked in the presence of another polymer. The constituent polymers in

IPN are not soluble in solvents. However, in semi-IPN, one of the polymers remains uncross-linked, which is interwoven with the cross-linked counter component, and the uncross-linked polymer can be partially extracted by suitable solvents [13]. IPN allows the exhibiting of two or more distinguishable properties rendered by the constituent polymers. A typical example is epoxy/polyurethane IPN, where epoxy offers strength and modulus, whereas the rubbery polyurethane counters the brittleness of epoxy [13]. IPNs are suitable for electrical and thermal conductivity, improving strength and toughness, stimuli-responsive applications, drug delivery, etc.

Another approach to making a hybrid composite is by developing an organic-inorganic matrix. The organic part, i.e., hydrocarbon polymers, offers processability, weight reduction, flexibility, and high material design versatility. At the same time, the inorganic component provides thermal and electrical conductivity, mechanical properties, and a unique combination of optical, catalytic, and magnetic properties. The inorganic parts comprise metals and ceramics, or both, and have macro- and nano-dimensions. The organic-inorganic hybrids are synthesized by mechanical blending, the sol-gel method, in-situ emulsion polymerization, supramolecular and coordination approaches, photopolymerization, etc. [11]. Among them, the sol-gel method and supramolecular self-assembly are universally used. The sol-gel method is an inorganic polymerization technique where metal alkoxide undergoes hydrolysis and condensation reactions [14]. The most studied system is a silica-based polymer hybrid. The hybrid network is formed by using organo alkoxysilane as a precursor where organic groups are introduced to the inorganic network by Si-C linkage. Swelling of metal alkoxide in the cross-linked, ionomeric, or crystalline polymer matrix followed by the sol-gel reaction is also a popular method of producing an in-situ organic-inorganic hybrid network [14]. On the other hand, the supramolecular approach of forming an organic-inorganic network is based on noncovalent interactions, which include metal ligands, H-bonding, and ionic bonding. The linear polymer-metal hybrid chain can be formed by including metal ions within the small organic molecular unit containing two end-ligating functionalities. When metal ions couple the homopolymers of different functionalities, it forms a block copolymer-metal hybrid. The inclusion of metals is also possible through side chain moieties instead of the backbone of the main chain.

1.4 Hybrid fiber composite

FRP composites (FRPs) are attractive due to their high specific strength, stiffness, corrosion resistance, and excellent fatigue properties. FRPs offer a plethora of design flexibilities by varying the fiber type, fiber length, orientation, stacking sequence, and preform types. The concept of developing a hybrid fiber composite containing more than one type of fiber provides solutions for cost-cutting, processability, improved mechanical behavior, and other multifunctional properties. In the case of continuous FRPs, hybridization in the filler system can be carried out in three ways: intrayarn, intralayer, and interlayer (Figure 1.1) [15]. The intra- or interlayer hybrid fiber composite can be made unidirectionally and multidirectionally. An example of a commercially available multidirectional hybrid fiber is the woven fabric containing carbon fiber in the weft direction and glass fiber in the warp direction. When two fibers are combined, the properties of the composite primarily depend on the fine dispersion of the fibers, characteristics of individual fibers, e.g., modulus, ultimate tensile strength, density, and elongation at break, and their volume fractions. In general, the tensile strength of a hybrid fiber composite follows the binary rule of mixture. The volume fraction of component fibers determines whether the failure of the composite will be associated with the high or low elongation plies or bundles. The thermal expansion coefficient of the individual fibers is also an important criterion for selecting multiple fiber systems. Due to the significant difference in the thermal expansion coefficient between carbon fiber and polymer fiber, their hybrid composite often suffers from residual thermal stresses [15]. Thermal stresses can be reduced by modifying the curing or molding cycle. Due to stringent environmental regulations, the usage of natural fiber-based biocomposites is continuously growing in various applications. Natural fibers are low-cost and lightweight, but their strength and modulus are not as high as synthetic fibers. Natural fibers, which are mainly cellulose-based, contain hydroxyl groups that swell in the presence of moisture, resulting in unsatisfactory properties. These drawbacks can be overcome by hybridizing with synthetic fibers, such as glass fiber, carbon fiber, basalt fiber, etc. (Figure 1.1). Innumerable studies have been reported with various natural fiber-based hybrid composites, e.g., coir, sisal, jute, bamboo, hemp, kenaf, sugar palm, banana, flax, PALF, etc. [16, 17]. Among the hybrid biocomposites, the most research has been conducted on glass fiber/cellulosic natural fiber composites [18]. Natural fibers are often subjected to surface treatment to improve the compatibility between the natural fibers, which are highly polar and nonpolar synthetic fibers and different polymer matrices. For example, alkaline treatment (treated with NaOH), silane treatment, acetylation of natural fiber, benzoylation treatment, maleated coupling agents, etc. [19].

The state-of-the-art nanoscale materials (e.g., graphene, carbon nanotubes, carbon dots, silver nanoparticles, silver nanowires, nanocellulose, MoS_2, MXene, layered double hydroxide (LDH), etc.) are promising for introducing multifunctional properties in hybrid nanocomposites due to their exceptional mechanical, electrical, chemical, optical, and other physical properties. Nanofillers are primarily divided into three categories based on their shape: particulate type (e.g., nanodots), one-dimensional (1D) (e.g., nanorods, nanofibers, nanowires, etc.), and two-dimensional (2D) (e.g., nanoplatelets, layered structures, nanoribbons, etc.). Nanofillers can be introduced to the matrix or grafted onto the fiber or fabrics in the hybrid nanocomposites. The nanofiller dispersion is crucial in this case as they tend to agglomerate due to their high surface energy. Dispersibility can be improved by sonication, adding dispersing agents, surface functionalization, in-situ synthesis, etc. Modifying the fiber surface with nanofillers like graphene, CNT, graphene nanoplatelets, silver nanowires, etc., has gained significant attention in developing advanced multifunctional composites. There are several articles on the grafting of carbon fibers with carbonaceous fillers (such as carbon nanotubes, graphene, and carbon blacks). These nanofillers are attached to the carbon fiber surface via pi-pi interactions, chemical bonding through surface functionalization, or coupling agents [20–24] (Figure 1.1). Another popular way of grafting the nanofiller onto the fiber surface is by synthesizing the nanomaterials in situ. Incorporation of the nanofiller results in forming an interconnected three-dimensional network, drastically improving conductivities, mechanical properties, fatigue properties, barrier properties, heat- and flame-resistant properties, etc.

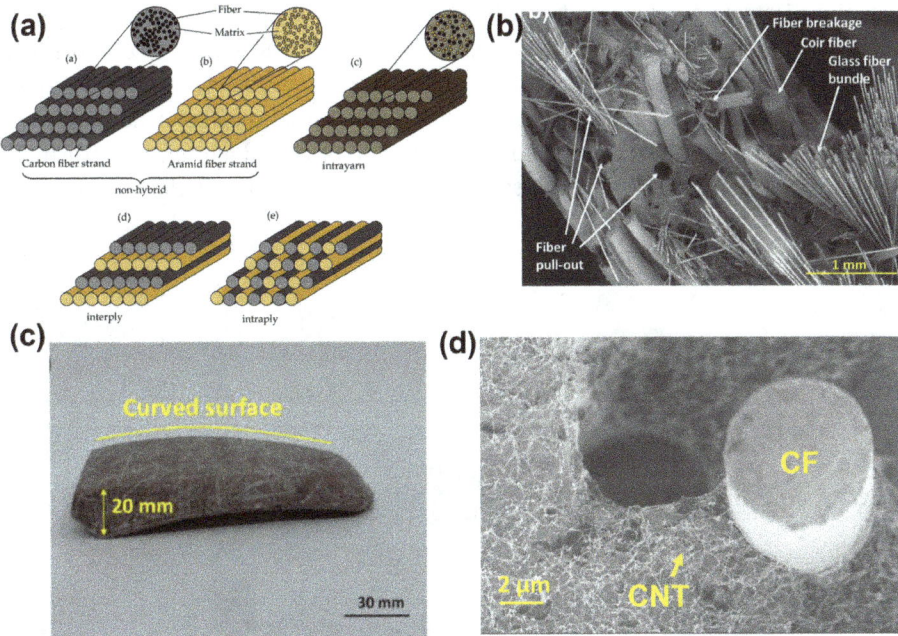

Figure 1.1: (a) Approaches to form hybrid fiber composites [25], (b-c) natural long coir fiber/GF composites [17], and (d) hybrid CNT/CF composites [23].

1.5 Hierarchical composites

The composites containing well-ordered structures with several length scales (generally three or more) mimicking nature are called hierarchical composites. It has been observed that the unique properties of biological materials are attained through the hierarchical structure. The best examples are bone, nacre, crab carapace, bamboo, tendon, and spider silk. These structures contain different levels of hierarchy, and each level is constructed by arranging building blocks in the form of staggered structures, bouligand structures, self-similar fiber bundles, concentric multi-walled cylinders, and gradient designs [26]. Based on the definition, nanohybrid composites also fall into this category. Hierarchical composite structures have shown immense potential in acquiring an excellent combination of properties by introducing multi-length scale-controlled complex designs. For example, the improvement of both modulus and impact properties can be achieved through hierarchical composite design for structural applications. To improve the mechanical properties, the staggered and bouligand structures are found to be most effective (Figure 1.2). The bouligand pattern is inspired by arthropods and fish scales, attributing a twisted stair-like stacked lamellar structure. The twisted structure has two distinct advantages in improving the reinforcement [26]. The first one is the introduction of in-plane isotropy through the twisted structure, which avoids weaknesses in specific directions, unlike the unidirectional fiber arrangement. The second one is the self-adaptive reorientation of the fibers toward the loading direction, delaying crack initiation and propagation compared to conventional composites. Inside bone, hard minerals and soft matrix are organized in the form of staggered brick-and-mortar wall structures, which are excellent in maintaining strength and toughness as per the tension-shear

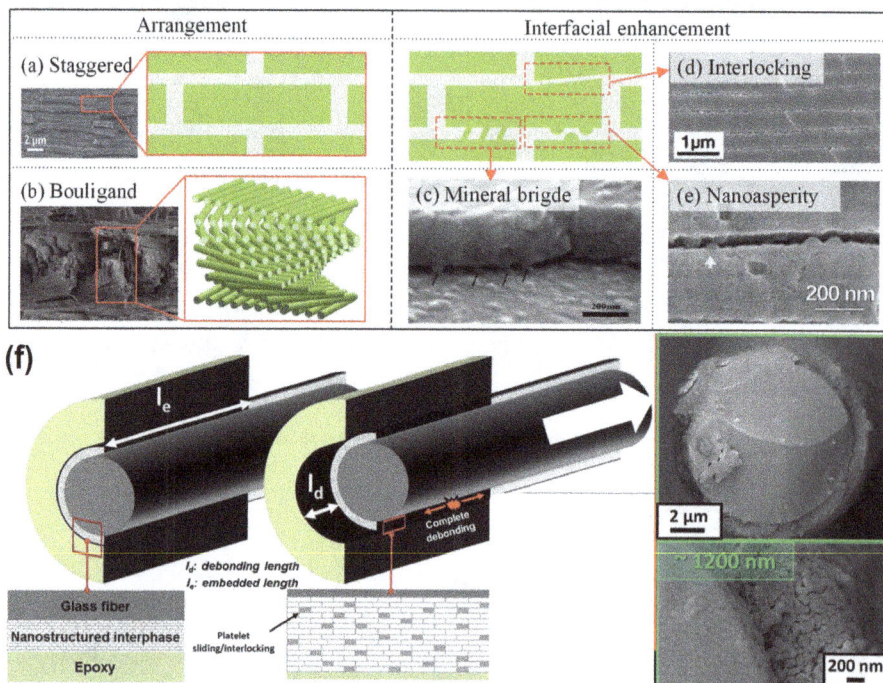

Figure 1.2: Hierarchical composites, (a–e) different biomimetic arrangements [26], (f) nanostructured layered double hydroxide (LDH) coating on GF [28].

chain model proposed by Gao [26, 27]. Nanostructured materials (e.g., nanotubes)-grafted fibers with stitched interlaminar morphology mimic the brick-and-mortar structure, and such morphology has proven to be efficient in improving the interfacial shear or the fracture toughness of the glass fiber-reinforced composite (Figure 1.2) [28].

1.6 Hybrid manufacturing

Over a long period, the conventional manufacturing techniques involved in the mass production of thermoplastic and thermoset composites include extrusion, pultrusion, compression molding (ECM) [29], injection molding (IM) [30], blow molding, hand layup, resin transfer molding, braiding processes [31], automatic tape layup (ATL) [32], filament winding [33], etc. The selection between various manufacturing methods for composite parts depends on the following factors: type of polymer matrix (thermoplastic or thermoset) and reinforcement, design complexity of the part, desired properties, production volume, and cost-effectiveness. Traditional manufacturing methods often encounter several challenges in achieving precise fiber orientation, uniform distribution and wetting, and complex geometry, alongside high tooling costs for intricate molds, resulting in overdesign or inefficiencies. For example, a typical extrusion or pultrusion process offers high production rates and is suitable for continuous production, making it ideal for manufacturing profiles or structures in large volumes. However, the major limitation lies in producing complex geometry. Compression molding is another widely used method for both thermoplastics and thermosets. However, the process lacks a high production rate, controlling fiber orientation and wetting of fibers, leading to resin-rich or fiber-rich areas within the final part [34–37]. Injection molding, on the other hand, provides high repeatability and precision for thermoplastic composites, enabling the manufacturing of complex geometries. Nonetheless, it also faces limitations in controlling fiber orientation, incomplete filling, and high tooling costs for complex molds [38]. Hand layup, resin transfer molding, vacuum-assisted resin transfer molding, etc., frequently used in manufacturing thermosets, often face challenges in fast production, complicated geometry, materials waste, etc. [39–41]. Braiding, an automated continuous process, offers good control over fiber orientation, maintaining an excellent strength-to-weight ratio and low labor costs; however, it is confined to tubular or near-tubular shapes, with challenges in achieving uniform resin impregnation [31]. The automatic tape layup or filament winding process also faces challenges in laying tapes in complex geometries and controlling resin distribution, potentially leading to variations in part quality [32, 33].

Hybrid manufacturing approaches by integrating multiple fabrication techniques render promising solutions to address these limitations and enhance composite manufacturing processes' efficiency, precision, and versatility [42–44]. For example, integrated injection-compression molding (ICM) requires less injection pressure compared to traditional injection molding, combined with the reduction in packing-induced anisotropy, lateral shrinkage, and residual strain (Figure 1.3) [43, 45]. However, ICM still needs high tooling costs for producing intricate parts and offers no control over fiber orientation. In this aspect, additive manufacturing (AM) or 3D printing of polymer composites offers several advantages, including enhanced optimal fiber orientation, cost-effective tooling, reduced material waste, and the ability to produce complex geometries with customized properties [46]. Additive manufacturing fabricates objects layer by layer from a design model created using computer-aided design (CAD) software. The model is divided into discrete slices specifying where the material will be deposited in each layer, and then the object is fabricated layer by layer. Each successive layer adheres securely to the one below it, thereby contributing to the formation of the complete 3D structure. Additive manufacturing can significantly align the fibers in the deposition direction due to the shear stresses generated during the material extrusion within the nozzle. The digital control enables precise manipulation of various parameters such as layer thickness, material composition, and fiber orientation, leading to enhanced customization and optimization of part properties. Despite these advantages, critical flaws associated with the AM process include excessive porosity within printed beads and suboptimal bead-to-bead interface undermining the mechanical properties of 3D-printed parts [47, 48]. Combining AM with conventional compression molding (AM-CM) offers a solution by consolidating the structure to reduce porosity [34]. The AM-CM process involves the fabrication of a preform by AM followed by compression molding to produce the final parts with preferred fiber orientation, high density, and high structural integrity (Figure 1.3). Similar to AM-CM, AM can indeed be combined with injection molding [49], automated fiber placement or tape layup [50–52], and filament winding [53] to enhance manufacturing processes and achieve specific goals. In a true sense, additive manufacturing-injection molding (AM-IM) follows the concept of over-molding by inserting a 3D-printed preform within the injection-molded parts [54]. The preform is usually fabricated with continuous fabric to provide structural strength, while the injection over-molded material offers functionality.

Meanwhile, combining additive manufacturing with automated fiber placement (AM-AFP) improves the geometric complexity of composite structures. AM-AFP encompasses two separate robots dedicated to AM and AFP processes, respectively, aiming to work together (Figure 1.3). AM robots fabricate the mold or structure on which the AFP robot lays the prepreg tows, targeting to improve the design and production workflow of composite parts, covering the development of an effective design environment for fiber layout and a robust machine control system [50]. Similarly, combining AM with filament winding (AM-FW) allows for creating customized mandrels, reinforcement structures, and internal features within composite parts, offering enhanced design flexibility and structural optimization. These hybrid approaches offer superior control over part design, production efficiency, and material utilization, catering to diverse industries and applications such as aerospace, automotive, sporting goods equipment, etc.

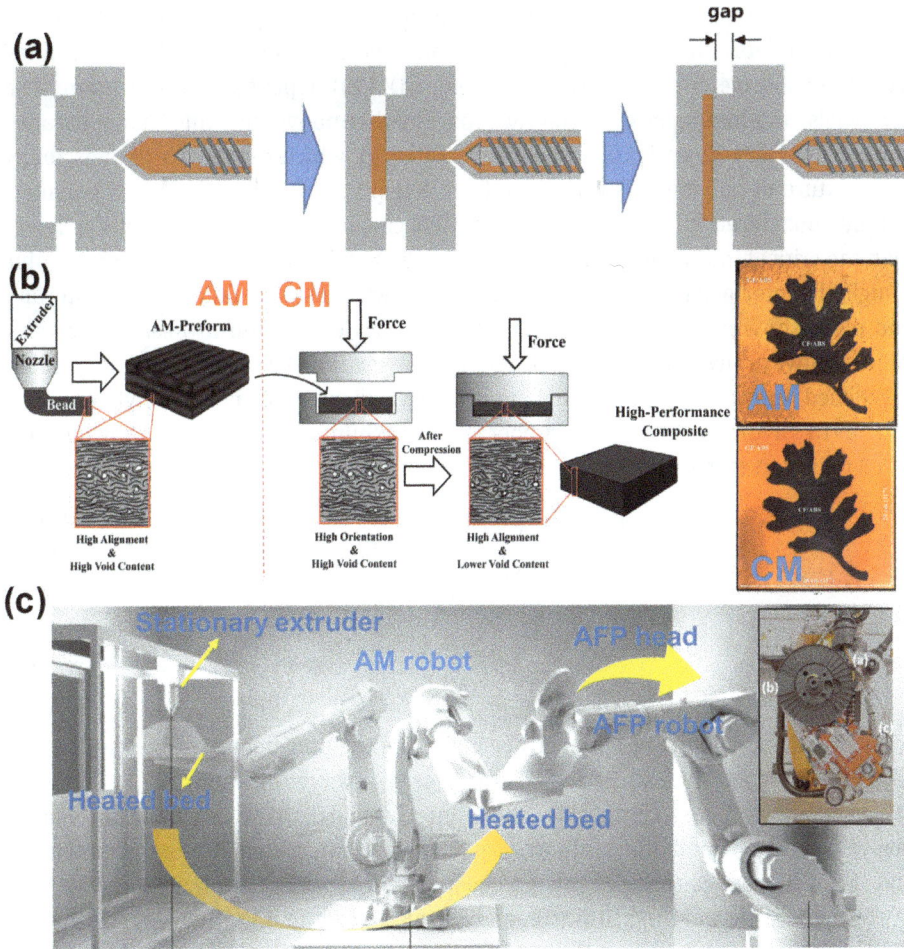

Figure 1.3: The schematic of different hybrid manufacturing processes: (a) ICM [43], (b) AM-CM, and (c) AM-AFP [50].

1.7 Properties of the hybrid composites

1.7.1 Interface properties and stress concentration

Composites are multiphase materials with distinct interfaces leading to the stress gradient between different constituent components. Favorable interaction between the interfaces is necessary to reduce the interfacial tension for suppressing stress concentration, uninterrupted electron and phonon conduction, minimizing interfacial polarization, etc. Solubility and interaction parameters are the thermodynamic properties that govern the miscibility between different polymers in hybrid matrix composites [55]. The Hildebrand solubility parameter (δ) governing the miscibility is obtained from the cohesive energy density (CED) (eq. 1.1). CED depends on electrostatic, Van der Waals, and hydrogen bond interactions. The thermodynamic interaction parameter, i.e., the Flory-Huggins interaction parameter (χ), which represents the miscibility of a mixture of polymers, is calculated from δ (eq. 1.2) [56]. The mixture of polymers will be considered miscible if the interaction parameter value of the mix is lower than the critical interaction parameter (χ_c) (eq. 1.3). If the interaction parameter value is high, the interfacial stress transfer will be restricted, causing failure at the interface and often showing lower strength, toughness, and elongation at break. A lower χ value is always preferred for exhibiting a stronger interface. For example, rubber-toughened thermoplastics, PP/ethylene propylene diene rubber, polyvinylidene fluoride (PVDF)/hydrogenated nitrile rubber (HNBR) exhibited low χ values, showing a delayed failure in tensile mode [57, 58]:

$$\delta = \sqrt{CED} \tag{1.1}$$

$$\chi = \frac{V_m}{RT}\left[(\delta_1 - \delta_2)^2\right] \tag{1.2}$$

$$\chi_c = \frac{1}{2}\left(\frac{1}{\sqrt{n_A}} + \frac{1}{\sqrt{n_B}}\right)^2 \tag{1.3}$$

where V_m is the volume of polymer per mole, R is the universal gas constant, T is the temperature in absolute scale, and n is the degree of polymerization.

In fiber-reinforced composites, sharp interfaces exist between fiber and matrix due to a local mismatch of stiffness, causing fiber-matrix debonding under tension and flex. Different coupling agents and coatings (fiber sizing) are applied to the fiber to strengthen the bonding. It has been observed that the nanofiller-grafted hybrid fiber composite exhibits better mechanical interlocking between fiber and matrix. Carbonaceous nanofillers, such as graphene, CNT, carbon nanofiber, etc., can act as bridging elements to improve load transfer in the case of carbon fiber-reinforced composite [59]. Bonding between nanofillers/CF, nanofiller/matrix, and CF/matrix determines the interfacial shear strength. Zhang et al. observed the filamentous bridging of carbon fibers by the bundle of CNTs [60]. Due to such bridging, additional energy was needed for crack propagation, and the crack primarily propagated through the matrix phase instead of the fiber-matrix interphase. Nanoclays, like montmorillonite, Cloisite, etc., are being used by several researchers to improve the interfacial shear strength of glass fiber-reinforced composites. The nanoclays remained at the fiber-matrix interface region, which exhibits a rough fracture surface, indicating stronger bonding between glass-fiber epoxy. Wither et al. reported a 10.6% and 7.9% improvement in tensile modulus and fatigue strength, respectively by adding oregano-modified Cloisite B [61]. Rafiq et al. showed improved impact resistance properties in hybrid nanoclay GFRP composites [62]. With the rise in stiffness upon incorporation of the nanoclay, the buckling of fibers is restricted, ensuring better fiber-matrix bonding. It also effectively blunts the crack propagation through matrix and interface regions where the resistance is less than that of the fiber, consequently delaying the failure.

Manufacturing of composites also governs the stress concentration and interface properties. Injection molding has been extensively utilized in automotive component manufacturing. However, the high injection pressure involved in this process leads to pressure-induced residual stresses [63]. In contrast, the hybrid ICM process requires lower injection pressure, minimizing the residual stresses [43]. Compared to injection molding and compression molding, the stress concentration is considerably higher for the parts manufactured by AM due to layer-by-layer deposition. Poor diffusion between the deposited layers, voids and uneven cooling associated with the additive manufacturing process result in weak interfaces, and stress concentration. AM followed by compression molding suppresses the porosity and promotes interface diffusion between the layers, improving the surface quality and reducing the stress concentration. In the AFP process, fiber tow gap is one of the prime contributors to the stress concentration, along with other defects such as overlapping, wrinkling, missing or twisted tow, upfolding, bridging, crowning, etc. The gap areas in the composite structures lead to plastic deformation of the matrix, which adversely impacts the mechanical properties, fatigue strength, and damage tolerance of the composites. Rakhshbahar et al. reported a potential solution to minimize the effect of fiber tow gap by post-manufacturing 3D printing with continuous carbon fiber-reinforced prepreg [51]. Besides, the fiber tow gap can be reduced by customizing the mold design for the AFP process, which can be easily attained by opting for the hybrid AM-AFP method.

1.7.2 Mechanical properties

Mechanical properties are the primary concern for structural composites where FRPs are exclusively used. In fiber-reinforced composites, although the fibers (or fabrics) control the mechanical properties, the matrix also contributes to the stiffness, especially in anisotropic composites, such as oriented FRPs, in the direction where the fibers are not aligned. The ratio of the transverse stiffness to the matrix stiffness determines the stress concentration [15]. A stiffer matrix produces higher shear stresses, generating more local stresses. The development of a hybrid matrix containing macro rubber particles has been used to suppress matrix cracking in many research articles [64, 65]. The rubber toughening mechanism can improve the fatigue and impact properties of inherently brittle epoxy. Turcsán and Mészáros reported the hybridization of two thermosetting resins, epoxy and vinyl ester, which formed an IPN structure and such an IPN structure enhanced the toughness of the CF/GF-reinforced composites [66]. Czigány et al. also noticed a similar observation when a hybrid epoxy and vinyl ester matrix was reinforced with a discontinuous basalt fiber mat [67].

The mechanical properties of glass/carbon fiber hybrid composites have been exhaustively investigated. In general, glass fiber has lower strength, Young's modulus, and Poisson's ratio compared to carbon fiber, whereas the cost of glass is significantly lower. Additionally, carbon fiber, being a linear elastic material, is highly sensitive to notches and does not provide any warning before final failure. Hybridized glass and carbon fiber showed synergistic mechanical properties as described by several researchers [15, 68, 69]. One of the reasons for this synergism is the pseudo-ductile behavior, which has been proven to improve notch sensitivity. Mineral basalt fiber-based hybrid fiber composites are well known for the economic optimization of high-cost FRPs, and these fibers also offer a higher modulus and strength compared to E-glass fiber [70]. A study showed that the mechanical properties did not change significantly with the increasing amount of cheap basalt fiber content in the CF/basalt hybrid composite (20–40 wt%) [70]. In another study, Cao et al. observed similar mechanical properties of hybrid carbon/glass fiber and carbon/basalt fiber-based composites [71]. Similarly, natural fiber-based biohybrid composites are also used to reduce costs along with an excellent strength-to-weight ratio [7]. Polymer fibers, such as aramid fiber, polyacrylate fiber, ultrahigh molecular weight polyethylene (UHMWPE), polybenzobisoxazole (PBO), etc., contain highly oriented polymer chains in the fiber directions, which leads to high longitudinal stiffness and strength. These fibers fibrillate in tension and yield micro kinks in compression, making them adaptive to absorb a large amount of energy. Hybrid polymer fibers combined with carbon fibers offer superior impact resistance compared to all-carbon fiber composites [15]. Recent studies revealed that incorporating metal fibers or metal wires improved the penetration impact resistance and notch sensitivity of the hybrid composites [72, 73]. Zinc oxide nanowires or nanorods provide superior vibration control for structural applications in combination with conventional FRPs [74]. Hybrid preform technology also showed a significant impact on the mechanical properties. One such example is the introduction of a thin ply of thickness below 100 µm into the continuous laminate structure. The thin ply inserts are susceptible to suppressing the unstable delamination, which avoids significant load drop during the breakage of the brittle components. This allows gradual load transfer from brittle to the more ductile component, leading to pseudo-ductility [75].

1.7.3 Functional properties

Apart from improving the mechanical properties, another aspect of hybridization is to append multifunctional properties to the composites. The hybridization process is highly effective in improving the electronic conductivity of polymer composites by forming a three-dimensional interconnecting network. Polymers are generally non-conductive except for a few intrinsically conducting polymers, e.g., polyaniline, polypyrrole, polyindole, polyacetylene, etc. Hence, the conductivity of the fiber-reinforced composite, especially in the case of short fiber, lies in the low range. The incorporation of conducting filler escalates electron conduction through the polymer matrix; however, the conduction is solely dependent on the filler dispersion. Metal wires, graphite powder, carbon black, etc., are examples of conducting macro fillers. A remarkable increase in conductivity can be achieved by dispersing the nano-structured fillers.

Anisotropic high-aspect nanofillers, such as nanotubes, nanofibers, nanoribbons, nanoplatelets, nanoflakes, etc., facilitate electronic conduction. Among the different carbonaceous fillers, single and multi-wall carbon nanotubes are the most promising for electronic conductivity due to their one-dimensional nanostructure and lower band gap [23, 76]. Several strategies are adopted to improve the dispersion, including surface functionalization, grafting, doping, etc. Graphene, due to its sp^2-hybridized hexagonal layer structures, can show electrical conductivity up to 20,000 $S.m^{-1}$ [77]. Single-layer and multilayer graphene are successfully used to make electrically conductive hybrid composites. To improve the exfoliation, functional groups are introduced to the graphene surface by oxidation reaction. The surface functional groups of graphene oxide, such as hydroxyl, carboxylate, epoxy, etc., are detrimental to electrical conductivity. So, a further reduction is needed to retain higher electrical conductivity. Different nitrogen-containing reducing agents help in the reduction of the graphene oxide as well as enhance the bonding between other fillers to form a strong conductive network [78, 79]. Other carbonaceous nanofillers like carbon nanofibers, carbon nanorods, and 3D graphene also showed significant improvement in the electrical conductivity of the fiber-reinforced composites. The recent discovery of 2D transition metal carbide/nitride/carbonitride named MXene has gained notable research interest for improving the electronic conduction in the composites [80]. MXene is unstable in heat and aerial environments. Thus, a strategy of blending with other fillers has been adopted by many researchers to form a hybrid composite that can offer high electrical conductivity suitable for use in high-performance applications such as smart fabrics, sensors, stretchable electronics, EMI shielding, etc. [81]. Different conductive metallic nanofillers, such as silver nanowires or nanoparticles containing hybrid composites, are also effective in attaining high electrical conductivity. The through-thickness conductivity is always challenging, even for the continuous carbon fiber-reinforced composites, as the interlaminar region is rich with the insulating polymer. Several studies showed that the vertically oriented nanofillers, such as CNT, carbon nanofibers, graphene, etc., grafted on the carbon fiber surface significantly improve the through-thickness conductivity. Instead of growing other nanofillers in the vertical direction, Kumar et al. have investigated the through-thickness conductivity of a composite composed of vertically oriented short carbon fibers in combination with conventional woven carbon fiber fabric [82]. They reported that the through-thickness conductivity increased by 800% with the interleaved vertical carbon fiber, significantly improving lightning strike protection.

The thermal conductivity of composites is another crucial property for many potential applications, including aerospace, automotive, electronic, energy storage, packaging, barrier, and high-pressure storage. Thermal conduction through the composite depends on electron and phonon conduction, crystalline structure, and interfacial thermal resistance between the phases caused by phonon mismatch [83]. Various inorganic ceramic fillers have inherently high thermal conductivity, which has been utilized to develop hybrid thermally conductive composites. Typical examples are barium titanate, boron nitride, alumina, titanium dioxide, zinc oxide, silicon carbide, and silicon nitride. Nanoscale-level inorganic filler can offer further improvement in conductivity. In the case of thermal conductivity, interfacial resistance has a significant impact. Thus, a marginal improvement in thermal conductivity is often observed even in the presence of a large quantity of highly thermally conductive fillers. Improper connectivity at the interfaces also induces resistance. The combination of isotropic and anisotropic hybrid filler systems has been proven to form a suitable interconnected 3D template that enhances thermal conduction [84]. TabkhPaz et al. reported a 290% increase in thermal conductivity by blending 3.1 vol% of 1D CNT into layered structured hexagonal boron nitride compared to the single type of filler [85]. Zhu et al. revealed that the incorporation of hBN of different sizes manifested a higher thermal conductivity than the single particle size at similar filler loading [86]. Hybridization using ceramic filler can improve the thermal conductivity of polymer composites without changing the electrical insulation properties [87].

Composites for high-temperature applications, electronic and electrical devices, and friction or dynamic applications require fire and flame resistance properties. When a composite material is exposed to a high heat flux, it suffers from softening and degradation of the matrix, matrix cracking, delamination of plies, debonding of matrix and fiber, and finally, loss of strength and stiffness [88]. Polymers are, in general, organic molecules containing hydrocarbons, most likely combustible; however, fibers, e.g., carbon fiber, glass fiber, basalt fibers, etc., are thermally more stable. On the other hand, biofiber-based composites are thermally labile, limiting their applications. The hybridization of bio-derived fiber with synthetic fibers or the incorporation of inorganic fillers is susceptible to improving the flame resistance properties [89]. The carbonaceous fiber or nanofiller also prevents material degradation by forming protective char, which restricts the transfer of combustible gases and heat. Incombustible inorganic fillers or mineral fillers can offer flame retardancy by reducing the total amount of fuel content and limiting oxygen diffusion. Additionally, the decomposition of some mineral fillers also produces water vapor through endothermic reactions, which help absorb the heat and hinder fire propagation, such as aluminum hydroxide and magnesium hydroxide [90]. Research showed that hybridization with inorganic nanofillers, e.g., nanoclay, nanosilica, layered double hydroxide, etc., is found to be promising to impart flame retardancy to the composite.

Composites for packaging applications and chemical and gas storage require moisture and gas barrier properties. Moisture sensitivity is one of the major drawbacks of bio-based composites. The incorporation of silica or clay particles can resist moisture. Two-dimensional fillers are most effective in improving the gas barrier properties by increasing the tortuosity of the matrix [91, 92]. The gas barrier property is the prime requirement for all composite-based compressed gas storage tanks. Current composite pressure vessels are built with carbon fiber composite. The diffusion of the gas through the composite layer suffers from delamination, blistering, and collapse of the composite structure. Insertion of a 2D layer structure nanofiller homogeneously distributed throughout the matrix allows the gases to diffuse following a long tortuous path, resulting in improved barrier performance.

There is growing interest in composites with magnetic and dielectric properties in a broad application area, which includes electronic applications, communication equipment, biomedical applications, sensors, energy storage applications, etc. Mixing magnetic powder, such as Fe_3O_4, $CoFe_2O_4$, strontium ferrite, nickel, etc., into composites enhances the electromagnetic wave absorption ability. Magnetic polymer composites can also offer other multifunctional properties such as shape memory, self-healing, sensing, etc. [93]. Dielectric ceramic filler-reinforced hybrid composites find applications in energy storage and energy conversion, EM wave absorption, integrated circuits, multilayer dielectric printed circuit boards, etc. Typical dielectric fillers include barium titanate, zirconium titanate, strontium titanate, titanium dioxide, aluminum oxide, zirconium oxide, etc. [94, 95]. The inclusion of magnetic particles can be done by melt or solution blending, sol-gel method, in-situ polymerization, chemical vapor deposition (CVD) method, and spin coating. Frequently used polymers in dielectric composites are aromatic polyimide, polynitrile, fluoropolymer, polyvinylidene fluoride, polyvinyl alcohol, polyethersulfone, epoxy, etc.

1.8 Applications

The aerospace and automotive industries are the most important application areas for hybrid composites. Aircraft need the highest level of concern regarding strength and weight. At the same time, functional properties like electrical conductivity, EMI shielding, flame resistance, thermal insulation, friction properties, and corrosion resistance are also crucial. Hybrid carbon and glass fiber epoxy composites are frequently used in rotor blades, wings, and fuselage. For the construction of the structural parts, continuous aramid, glass, and carbon fiber-reinforced epoxy aluminum polymer-metal laminates are the most desired materials [96]. In the twenty-first century, environmental awareness has resulted in the higher usage of natural fiber-based biohybrid composites in aviation industries. Hari et al. showed that kenaf fiber, among different natural fibers, exhibited the highest mechanical properties, which were comparable to glass fiber [97, 98]. The hybridization of glass-kenaf fiber can provide improved mechanical and rain erosion properties. Natural fibers, e.g., flax, hemp, etc.,-based hybrid composites are also studied for designing the aircraft interior body parts, such as ceiling and sidewall panels, insulation blankets, etc. Automobile industries utilize hybrid fiber composites containing a mixture of glass, carbon, and basalt fiber in several interior and exterior body parts. Different thermoplastic elastomeric blends and block copolymers, such as PP/ethylene propylene rubber (EPR), PP/EPDM, polystyrene-butadiene-styrene block copolymer, thermoplastic polyurethane, etc., are extensively used. Many renowned OEMs, such as Mercedes, Volkswagen, Audi, Daimler-Chrysler, and Opel-GM in the passenger car segments, already use natural bio-derived fibers combined with synthetic fiber. For example, Mercedes E-Class's panels, external underbody panels, and Volkswagen's door structure comprise phenol-formaldehyde/flax/glass fiber hybrid composites. The incorporation of natural fiber can offer up to a 15% reduction in weight as compared to glass fiber composite [7]. In Formula 1 cars, the hybrid of carbon/aramid fiber is mandatory in some of the body parts to protect the driver from impact and fragments in the event of a crash [15]. Multiscale hybrid nanocomposites containing nanofillers, like nanoclay, calcium carbonate nanoparticle, CNT, graphene, etc., show significant improvement in modulus and dimensional stability, heat distortion temperature, scratch and mar resistance, and thermal and electrical conductivity.

In the wind industry, the purpose of adopting hybrid composites is to reduce costs while maintaining mechanical strength. The current state of the art is glass fiber-reinforced composites in load-carrying beams and aerofoils [15]. For larger blades, hybrid glass and carbon fiber have shown several benefits: the moisture sensitivity and lower modulus of glass fiber can be mitigated by carbon fiber, while glass fiber not only reduces costs but also helps in lowering stress concentration, and hindering delamination [99]. Hybrid basalt fiber-reinforced composites exhibit advantages by reducing cost and weight while showing higher stiffness and strength than E-glass fiber [100]. To further improve the properties, many researchers are investigating new hybrid matrices containing epoxy and mullite-like crystals [101]. Biohybrid composites containing flax, hemp, jute fibers, etc., are promising for cost reduction in short wind blades [102, 103].

In marine applications, composites are used in hulls, bearings, propellers, hatch covers, exhausts, topside structures, railings, vessels of all types, valves, and other subsea structures. Fiberglass-reinforced unsaturated polyester resin has been broadly used since the beginning. Hybridizing glass fiber with carbon fiber, basalt fiber, and other natural fibers provides several benefits, including weight, strength, cost, sustainability, and recyclability. Their use is now significantly increasing.

Hybrid nanostructured composites find applications in smart fabrics, sensors, energy storage, supercapacitors, biomedical devices, waste-water treatment, protective coatings, and many other diverse fields. Conducting nanofillers, such as graphene, CNT, MoS_2, etc., show promising results in wearable sensors for monitoring body temperature, heart rate, perspiration level, physical motion, etc. Hybrid nanocomposites are successfully tested in proton exchange membranes for fuel cell applications to improve the stability, conductivity, and cost-effectiveness of Nafion-based membranes conventionally used in this application [104]. Modifying Nafion with aromatic sulfonated polymers, e.g., poly(ether ether ketone) (PEEK), poly(ether sulfone) (PES), polybenzimidazole

(PBI), etc., in combination with inorganic nanofillers has attracted interest. The organic-inorganic hybrid nanocomposites containing hydrophilic nanoparticles (titanium dioxide, zirconium dioxide, sulfonated silica, nanoclay, etc.) forming an interconnecting channel in the hydrophobic polymer domain have been found to improve proton conductivities [104]. Organic-inorganic hybrid nanocomposites possess advantages over ceramic composites in dielectric applications due to easy processability and flexibility. Nanoclay-based hybrid composites are preferred for protective coatings with high thermal conductivity, scratch resistance, barrier properties, etc. Hybrid nanocomposites are also emerging in biosensors, controlled drug delivery, and other biomedical applications. In the era of intelligent, automated technology, hybrid materials play a major role in fulfilling the demand for advanced functional composites. Research on hybrid composites is exponentially gaining interest in several scientific communities, including chemists, physicists, biologists, materials scientists, and mechanical and other engineering fields. Many hybrid composite materials and manufacturing processes have already appeared as prototypes or commercial products, but this is just the tip of the iceberg. With more understanding and knowledge, many more hybrid composite materials and manufacturing methods will be seen in the upcoming years in different sectors, reinforcing their significance in modern life.

References

[1] C.-S. Kim, C. Randow, T. Sano Hybrid and Hierarchical Composite Materials, 1st 2015. ed., Springer International Publishing, Cham, 2015.

[2] M. Nanko Definitions and categories of hybrid materials, Advances in Technology of Materials and Materials Processing, 2009, 11: 1–8.

[3] Z. Zhu, V.G. Dhokia, A. Nassehi, S.T. Newman A review of hybrid manufacturing processes – State of the art and future perspectives, International Journal of Computer Integrated Manufacturing, 2013, 26(7): 596–615.

[4] J. Min, J. Hu, C. Sun, H. Wan, P. Liao, H. Teng, J. Lin Fabrication processes of metal-fiber reinforced polymer hybrid components: A review, Advanced Composites and Hybrid Materials, 2022, 5(2): 651–678.

[5] S. Li, P. Cheng, S. Ahzi, Y. Peng, K. Wang, F. Chinesta, J.P.M. Correia Advances in hybrid fibers reinforced polymer-based composites prepared by FDM: A review on mechanical properties and prospects, Composites Communications, 2023, 40: 101592.

[6] K. Jarukumjorn, N. Suppakarn Effect of glass fiber hybridization on properties of sisal fiber–polypropylene composites, Composites Part B: Engineering, 2009, 40(7): 623–627.

[7] M.J. Suriani, R.A. Ilyas, M.Y.M. Zuhri, A. Khalina, M.T.H. Sultan, S.M. Sapuan, C.M. Ruzaidi, F.N. Wan, F. Zulkifli, M.M. Harussani, M.A. Azman, F.S.M. Radzi, S. Sharma Critical review of natural fiber reinforced hybrid composites: Processing, properties, applications and cost, Polymers, 2021, 13(20): 3514.

[8] Z. Ying-Chen, W. Hong-Yan, Q. Yi-Ping Morphology and properties of hybrid composites based on polypropylene/polylactic acid blend and bamboo fiber, Bioresource Technology, 2010, 101(20): 7944–7950.

[9]   A. Ajji, L.A. Utracki Interphase and compatibilization of polymer blends, Polymer Engineering and Science, 1996, 36(12): 1574–1585.

[10] Frontmatter. Principles of Polymerization, John Wiley & Sons, Inc., Hoboken, New Jersey, 2004, pp. i–xxiv.

[11] H. Dau, G.R. Jones, E. Tsogtgerel, D. Nguyen, A. Keyes, Y.-S. Liu, H. Rauf, E. Ordonez, V. Puchelle, H. Basbug Alhan, C. Zhao, E. Harth Linear Block Copolymer Synthesis, Chemical Reviews, 2022, 122(18): 14471–14553.

[12] M.K. Purkait, M.K. Sinha, P. Mondal, R. Singh Chapter 3 – Temperature-responsive membranes, in: M.K. Purkait, M.K. Sinha, P. Mondal, R. Singh, (Eds.), Interface Science and Technology, Elsevier, 2018, pp. 67–113.

[13] S. Saha, W. Son, N.H. Kim, J.H. Lee Fabrication of impermeable dense architecture containing covalently stitched graphene oxide/boron nitride hybrid nanofiller reinforced semi-interpenetrating network for hydrogen gas barrier applications, Journal of Materials Chemistry A, 2022, 10(8): 4376–4391.

[14] J. Wen, G.L. Wilkes Organic/inorganic hybrid network materials by the sol–gel approach, Chemistry of Materials, 1996, 8(8): 1667–1681.

[15] Y. Swolfs, I. Verpoest, L. Gorbatikh Recent advances in fibre-hybrid composites: Materials selection, opportunities and applications, International Materials Reviews, 2019, 64(4): 181–215.

[16] S.O. Ismail, E. Akpan, H.N. Dhakal Review on natural plant fibres and their hybrid composites for structural applications: Recent trends and future perspectives, Composites Part C: Open Access, 2022, 9: 100322.

[17] S. Wasti, A.M. Hubbard, C.M. Clarkson, E. Johnston, H. Tekinalp, S. Ozcan, U. Vaidya Long coir and glass fiber-reinforced polypropylene hybrid composites prepared via wet-laid technique, Composites Part C: Open Access, 2024, 14: 100445.

[18] M.Ö. Seydibeyoğlu, A. Dogru, J. Wang, M. Rencheck, Y. Han, L. Wang, E.A. Seydibeyoğlu, X. Zhao, K. Ong, J.A. Shatkin, S. Shams Es-haghi, S. Bhandari, S. Ozcan, D.J. Gardner Review on Hybrid Reinforced Polymer Matrix Composites with Nanocellulose, Nanomaterials, and Other Fibers, Polymers, 2023, 15(4): 984.

[19] X. Li, L.G. Tabil, S. Panigrahi Chemical treatments of natural fiber for use in natural fiber-reinforced composites: A review, Journal of Polymers and the Environment, 2007, 15(1): 25–33.

[20] F. An, C. Lu, Y. Li, J. Guo, X. Lu, H. Lu, S. He, Y. Yang Preparation and characterization of carbon nanotube-hybridized carbon fiber to reinforce epoxy composite, Materials and Design, 2012, 33: 197–202.

[21] P.K. Biswas, O. Omole, G. Peterson, E. Cumbo, M. Agarwal, H. Dalir Carbon and cellulose based nanofillers reinforcement to strengthen carbon fiber-epoxy composites: Processing, characterizations, and applications, Frontiers in Materials, 2023, 9: 01–24.

[22] L. Chen, H. Jin, Z. Xu, M. Shan, X. Tian, C. Yang, Z. Wang, B. Cheng A design of gradient interphase reinforced by silanized graphene oxide and its effect on carbon fiber/epoxy interface, Materials Chemistry and Physics, 2014, 145(1): 186–196.

[23] S. Saha, V. Kumar, M.L. Rencheck, H. Tekinalp, B. Knouff, P. Blanchard, J. Yoon, K. Copenhaver, A.A. Hassen, H. Wang, S.M. Mahurin, K. Jayanthi, V. Kunc Development of multifunctional nylon 6,6-based nanocomposites with high electrical and thermal conductivities by scalable melt and dry blending methods for automotive applications, Materials Today Communications, 2024, 38: 107657.

[24] A.K. Pathak, M. Borah, A. Gupta, T. Yokozeki, S.R. Dhakate Improved mechanical properties of carbon fiber/graphene oxide-epoxy hybrid composites, Composites Science and Technology, 2016, 135: 28–38.

[25] T. Heitkamp, S. Girnth, S. Kuschmitz, G. Klawitter, N. Waldt, T. Vietor Continuous fiber-reinforced material extrusion with hybrid composites of carbon and aramid fibers, Applied Sciences, 2022, 12(17): 8830.

[26] Y. Chen, Y. Ma, Q. Yin, F. Pan, C. Cui, Z. Zhang, B. Liu Advances in mechanics of hierarchical composite materials, Composites Science and Technology, 2021, 214: 108970.

[27] H. Gao Application of fracture mechanics concepts to hierarchical biomechanics of bone and bone-like materials, International Journal of Fracture, 2006, 138: 101–137.

[28] F. De Luca, G. Sernicola, M.S.P. Shaffer, A. Bismarck "Brick-and-mortar" nanostructured interphase for glass-fiber-reinforced polymer composites, ACS Applied Materials and Interfaces, 2018, 10(8): 7352–7361.

[29] J. Jaafar, J.P. Siregar, C. Tezara, M.H.M. Hamdan, T. Rihayat A review of important considerations in the compression molding process of short natural fiber composites, The International Journal of Advanced Manufacturing Technology, 2019, 105(7): 3437–3450.

[30] M.S. Rabbi, T. Islam, G.M.S. Islam Injection-molded natural fiber-reinforced polymer composites–a review, International Journal of Mechanical and Materials Engineering, 2021, 16(1): 15.

[31] X. Sun, L.F. Kawashita, T. Wollmann, S. Spitzer, A. Langkamp, M. Gude Experimental and numerical studies on the braiding of carbon fibres over structured end-fittings for the design and manufacture of high performance hybrid shafts, Production Engineering, 2018, 12(2): 215–228.

[32] H.B. Olsen, J.J. Craig, Automated composite tape lay-up using robotic devices, Proceedings IEEE International Conference on Robotics and Automation, 1993, pp. 291–297

[33] S. Misri, M.R. Ishak, S.M. Sapuan, Z. Leman Filament winding process for kenaf fibre reinforced polymer composites, in: M.S. Salit, M. Jawaid, N.B. Yusoff, M.E. Hoque, (Eds.), Manufacturing of Natural Fibre Reinforced Polymer Composites, Springer International Publishing, Cham, 2015, pp. 369–383.

[34] V. Kumar, S.P. Alwekar, V. Kunc, E. Cakmak, V. Kishore, T. Smith, J. Lindahl, U. Vaidya, C. Blue, M. Theodore, S. Kim, A.A. Hassen High-performance molded composites using additively manufactured preforms with controlled fiber and pore morphology, Additive Manufacturing, 2021, 37: 101733.

[35] B. Šeta, M.T. Mollah, V. Kumar, D.K. Pokkalla, S. Kim, A.A. Hassen, J. Spangenberg, Modelling Fiber Orientation During Additive Manufacturing-Compression Molding Processes, (2022).

[36] P. Feraboli, T. Cleveland, M. Ciccu, P. Stickler, L. DeOto Defect and damage analysis of advanced discontinuous carbon/epoxy composite materials, Composites Part A, Applied Science and Manufacturing, 2010, 41(7): 888–901.

[37] B. Landry, P. Hubert Experimental study of defect formation during processing of randomly-oriented strand carbon/PEEK composites, Composites Part A, Applied Science and Manufacturing, 2015, 77: 301–309.

[38] C. Bivens, A. Wood, D. Ruble, M. Rangapuram, S.K. Dasari, K. Chandrashekhara, J. DeGrange Additively manufactured carbon fiber- reinforced thermoplastic composite mold plates for injection molding process, Applied Composite Materials, 2023, 30(5): 1569–1586.

[39] B. Karaçor, M. Özcanlı, H. Serin The influence of hybridization and different manufacturing methods on the mechanical properties of the composites reinforced with basalt, jute, and flax fibers, Iranian Polymer Journal, 2024, 33(3): 289–304.

[40] D. Cairns, J. Skramstad, J. Mandell, Evaluation of hand lay-up and resin transfer molding in composite wind turbine blade structures, 2001.

[41] V.R. Tamakuwala Manufacturing of fiber reinforced polymer by using VARTM process: A review, Materials Today: Proceedings, 2021, 44: 987–993.

[42] S.-J. Joo, M.-H. Yu, W. Seock Kim, J.-W. Lee, H.-S. Kim Design and manufacture of automotive composite front bumper assemble component considering interfacial bond characteristics between over-molded chopped glass fiber polypropylene and continuous glass fiber polypropylene composite, Composite Structures, 2020, 236: 111849.

[43] Y.I. Kwon, E. Lim, Y.S. Song Simulation of injection-compression molding for thin and large battery housing, Current Applied Physics, 2018, 18(11): 1451–1457.

[44] A. Siddiq, A.R. Kennedy Compression moulding and injection over moulding of porous PEEK components, Journal of the Mechanical Behavior of Biomedical Materials, 2020, 111: 103996.

[45] W.-S. Guan, H.-X. Huang Back melt flow in injection–compression molding: Effect on part thickness distribution, International Communications in Heat and Mass Transfer, 2012, 39(6): 792–797.

[46] S. Salifu, O. Ogunbiyi, P.A. Olubambi Potentials and challenges of additive manufacturing techniques in the fabrication of polymer composites, The International Journal of Advanced Manufacturing Technology, 2022, 122(2): 577–600.

[47] H.L. Tekinalp, V. Kunc, G.M. Velez-Garcia, C.E. Duty, L.J. Love, A.K. Naskar, C.A. Blue, S. Ozcan Highly oriented carbon fiber–polymer composites via additive manufacturing, Composites Science and Technology, 2014, 105: 144–150.

[48] C. Duty, J. Failla, S. Kim, T. Smith, J. Lindahl, V. Kunc Z-Pinning approach for 3D printing mechanically isotropic materials, Additive Manufacturing, 2019, 27: 175–184.

[49] D. Popescu, A. Zapciu, C. Tarba, D. Laptoiu Fast production of customized three-dimensional-printed hand splints, Rapid Prototyping Journal, 2019, 26: 134–144.

[50] F. Raspall, R. Velu, N.M. Vaheed Fabrication of complex 3D composites by fusing automated fiber placement (AFP) and additive manufacturing (AM) technologies, Advanced Manufacturing, Polymer and Composites Science, 2019, 5(1): 6–16.

[51] M. Rakhshbahar, M. Sinapius A novel approach: Combination of automated fiber placement (AFP) and additive layer manufacturing (ALM, Journal of Composites Science, 2018, 2(3): 42.

[52] H. Janssen, T. Peters, C. Brecher Efficient production of tailored structural thermoplastic composite parts by combining tape placement and 3D printing, Procedia CIRP, 2017, 66: 91–95.

[53] Y.D. Boon, S.C. Joshi, S.K. Bhudolia Review: Filament winding and automated fiber placement with in situ consolidation for fiber reinforced thermoplastic polymer composites, Polymers, 2021, 13(12): 1951 (01–29).

[54] R. Boros, P. Kannan Rajamani, J.G. Kovacs Combination of 3D printing and injection molding: Overmolding and overprinting, Express Polymer Letters, 2019, 13: 889–897.

[55] S. Saha, A.K. Bhowmick Computer aided simulation of thermoplastic elastomer from poly (vinylidene fluoride)/hydrogenated nitrile rubber blend and its experimental verification, Polymer, 2017, 112: 402–413.

[56] H. Lu, S. Du A phenomenological thermodynamic model for the chemo-responsive shape memory effect in polymers based on Flory–Huggins solution theory, Polymer Chemistry, 2014, 5(4): 1155–1162.

[57] P. Chakraborty, A. Ganguly, S. Mitra, A.K. Bhowmick Influence of phase modifiers on morphology and properties of thermoplastic elastomers prepared from ethylene propylene diene rubber and isotactic polypropylene, Polymer Engineering and Science, 2008, 48(3): 477–489.

[58] S. Saha, A.K. Bhowmick SMART thermoplastic elastomers with high extensibility from poly (vinylidene fluoride) and hydrogenated nitrile rubber: Processing–structure–property relationship, Rubber Chemistry and Technology, 2018, 91(1): 268–295.

[59] S. De, A.O. Fulmali, K.C. Nuli, R.K. Prusty, B.G. Prusty, B.C. Ray Improving delamination resistance of carbon fiber reinforced polymeric composite by interface engineering using carbonaceous nanofillers through electrophoretic deposition: An assessment at different in-service temperatures, Journal of Applied Polymer Science, 2021, 138(15): 50208.

[60] L. Zhang, N. De Greef, G. Kalinka, B. Van Bilzen, J.-P. Locquet, I. Verpoest, J.W. Seo Carbon nanotube-grafted carbon fiber polymer composites: Damage characterization on the micro-scale, Composites Part B: Engineering, 2017, 126: 202–210.

[61] G.J. Withers, Y. Yu, V.N. Khabashesku, L. Cercone, V.G. Hadjiev, J.M. Souza, D.C. Davis Improved mechanical properties of an epoxy glass–fiber composite reinforced with surface organomodified nanoclays, Composites Part B: Engineering, 2015, 72: 175–182.

[62] A. Rafiq, N. Merah, R. Boukhili, M. Al-Qadhi Impact resistance of hybrid glass fiber reinforced epoxy/nanoclay composite, Polymer Testing, 2017, 57: 1–11.

[63] C. Fernandes, A.J. Pontes, J.C. Viana, A. Gaspar-Cunha Modeling and Optimization of the Injection-Molding Process: A Review, Advances in Polymer Technology, 2018, 37(2): 429–449.

[64] C.M. Manjunatha, A.C. Taylor, A.J. Kinloch, S. Sprenger The effect of rubber micro-particles and silica nanoparticles on the tensile fatigue behaviour of a glass-fibre epoxy composite, Journal of Materials Science, 2009, 44(1): 342–345.

[65] M.R. Ricciardi, I. Papa, A. Langella, T. Langella, V. Lopresto, V. Antonucci Mechanical properties of glass fibre composites based on nitrile rubber toughened modified epoxy resin, Composites Part B: Engineering, 2018, 139: 259–267.

[66] T. Turcsan, L. Meszaros Mechanical performance of hybrid thermoset composites: Effects of matrix and reinforcement hybridization, Composites Science and Technology, 2017, 141: 32–39.

[67] T. Czigány, K. Pölöskei, J. Karger-Kocsis Fracture and failure behavior of basalt fiber mat-reinforced vinylester/epoxy hybrid resins as a function of resin composition and fiber surface treatment, Journal of Materials Science, 2005, 40(21): 5609–5618.

[68] X. Yan, S. Cao Structure and interfacial shear strength of polypropylene-glass fiber/carbon fiber hybrid composites fabricated by direct fiber feeding injection molding, Composite Structures, 2018, 185: 362–372.

[69] P.M. Bhagwat, M. Ramachandran, P. Raichurkar Mechanical Properties of Hybrid Glass/Carbon Fiber Reinforced Epoxy Composites, Materials Today: Proceedings, 2017, 4(8): 7375–7380.

[70] V. Fiore, T. Scalici, G. Di Bella, A. Valenza A review on basalt fibre and its composites, Composites Part B: Engineering, 2015, 74: 74–94.

[71] S. Cao, Z. Wu, X. Wang Tensile properties of CFRP and hybrid FRP composites at elevated temperatures, Journal of Composite Materials, 2009, 43(4): 315–330.

[72] B. Hannemann, S. Backe, S. Schmeer, F. Balle, U.P. Breuer, J. Schuster Hybridisation of CFRP by the use of continuous metal fibres (MCFRP) for damage tolerant and electrically conductive lightweight structures, Composite Structures, 2017, 172: 374–382.

[73] Y. Swolfs, P. De Cuyper, M.G. Callens, I. Verpoest, L. Gorbatikh Hybridisation of two ductile materials – Steel fibre and self-reinforced polypropylene composites, Composites Part A, Applied Science and Manufacturing, 2017, 100: 48–54.

[74] M.H. Malakooti, H.-S. Hwang, H.A. Sodano Morphology-Controlled ZnO Nanowire Arrays for Tailored Hybrid Composites with High Damping, ACS Applied Materials and Interfaces, 2015, 7(1): 332–339.

[75] G. Czél, M. Jalalvand, M.R. Wisnom Design and characterisation of advanced pseudo-ductile unidirectional thin-ply carbon/epoxy–glass/epoxy hybrid composites, Composite Structures, 2016, 143: 362–370.

[76] M.J. O'Connell Carbon Nanotubes: Properties and Applications, CRC Press, Boca Raton, Florida, 2018.

[77] X. Huang, X. Qi, F. Boey, H. Zhang Graphene-based composites, Chemical Society Reviews, 2012, 41(2): 666–686.

[78] C.K. Chua, M. Pumera Chemical reduction of graphene oxide: A synthetic chemistry viewpoint, Chemical Society Reviews, 2014, 43(1): 291–312.

[79] M. Kotal, A.K. Bhowmick Multifunctional hybrid materials based on carbon nanotube chemically bonded to reduced graphene oxide, The Journal of Physical Chemistry C, 2013, 117(48): 25865–25875.

[80] Y. Zeng, W. Wu Synthesis of 2D Ti3C2Tx MXene and MXene-based composites for flexible strain and pressure sensors, Nanoscale Horizons, 2021, 6(11): 893–906.

[81] H. Saini, N. Srinivasan, V. Šedajová, M. Majumder, D.P. Dubal, M. Otyepka, R. Zbořil, N. Kurra, R.A. Fischer, K. Jayaramulu Emerging MXene@Metal–organic framework hybrids: design strategies toward versatile applications, ACS Nano, 2021, 15(12): 18742–18776.

[82] V. Kumar, W. Lin, Y. Wang, R. Spencer, S. Saha, C. Park, P. Yeole, N.S. Hmeidat, C. Herring, M.L. Rencheck, D.K. Pokkalla, A.A. Hassen, M. Theodore, U. Vaidya, V. Kunc Enhanced through-thickness electrical conductivity and lightning strike damage response of interleaved vertically aligned short carbon fiber composites, Composites Part B: Engineering, 2023, 253: 110535.

[83] Z. Han, A. Fina Thermal conductivity of carbon nanotubes and their polymer nanocomposites: A review, Progress in Polymer Science, 2011, 36(7): 914–944.

[84] J. Yang, X. Shen, W. Yang, J.K. Kim Templating strategies for 3D-structured thermally conductive composites: Recent advances and thermal energy applications, Progress in Materials Science, 2023, 133: 101054.

[85] M. TabkhPaz, S. Shajari, M. Mahmoodi, D.-Y. Park, H. Suresh, S.S. Park Thermal conductivity of carbon nanotube and hexagonal boron nitride polymer composites, Composites Part B: Engineering, 2016, 100: 19–30.

[86] B.L. Zhu, J. Ma, J. Wu, K.C. Yung, C.S. Xie Study on the properties of the epoxy-matrix composites filled with thermally conductive AlN and BN ceramic particles, Journal of Applied Polymer Science, 2010, 118(5): 2754–2764.

[87] L. Zhang, H. Deng, Q. Fu Recent progress on thermal conductive and electrical insulating polymer composites, Composites Communications, 2018, 8: 74–82.

[88] X.-H. Shi, X.-L. Li, Y.-M. Li, Z. Li, D.-Y. Wang Flame-retardant strategy and mechanism of fiber reinforced polymeric composite: A review, Composites Part B: Engineering, 2022, 233: 109663.

[89] M. Bahrami, J. Abenojar, M.Á. Martínez Recent Progress in Hybrid Biocomposites: Mechanical Properties, Water Absorption, and Flame Retardancy, Materials, 2020, 13(22): 5145.

[90] T.R. Hull, A. Witkowski, L. Hollingbery Fire retardant action of mineral fillers, Polymer Degradation and Stability, 2011, 96(8): 1462–1469.

[91] K. Majeed, M. Jawaid, A. Hassan, A. Abu Bakar, H.P.S. Abdul Khalil, A.A. Salema, I. Inuwa Potential materials for food packaging from nanoclay/natural fibres filled hybrid composites, Materials and Design, 1980–2015, 46(2013): 391–410.

[92] O.B. Seo, S. Saha, N.H. Kim, J.H. Lee Preparation of functionalized MXene-stitched-graphene oxide/poly (ethylene-co-acrylic acid) nanocomposite with enhanced hydrogen gas barrier properties, Journal of Membrane Science, 2021, 640: 119839.

[93] Y. Fu, Z. Wei, Z. Wan, Y. Tian, Z. Zhao, L. Yang, G. Qi, G. Zhao Recent process of multimode stimuli-responsive flexible composites based on magnetic particles filled polymers: Characteristics, mechanism and applications, Composites Part A: Applied Science and Manufacturing, 2022, 163: 107215.

[94] H.P.P.V. Shanmugasundram, E. Jayamani, K.H. Soon A comprehensive review on dielectric composites: Classification of dielectric composites, Renewable and Sustainable Energy Reviews, 2022, 157: 112075.

[95] S. Saha, A.K. Bhowmick, A. Kumar, K. Patra, P.-J. Cottinet, K. Thetpraphi Polyvinylidene fluoride/hydrogenated nitrile rubber-based flexible electroactive polymer blend and its nanocomposites with improved actuated strain: Characterization and analysis of electrostrictive behavior, Industrial and Engineering Chemistry Research, 2020, 59(8): 3413–3424.

[96] E.C. Botelho, R.A. Silva, L.C. Pardini, M.C. Rezende A review on the development and properties of continuous fiber/epoxy/aluminum hybrid composites for aircraft structures, Materials Research, 2006, 9: 247–256.

[97] M.Y. Haris, D. Laila, E.S. Zainudin, F. Mustapha, R. Zahari, Z. Halim Preliminary review of biocomposites materials for aircraft radome application, Key Engineering Materials, 2011, 471–472: 563–567.

[98] N.J. Arockiam, M. Jawaid, N. Saba 6 – Sustainable bio composites for aircraft components, in: M. Jawaid, M. Thariq, (Eds.), Sustainable Composites for Aerospace Applications, Woodhead Publishing, 2018, pp. 109–123.

[99] Y. Swolfs Perspective for fibre-hybrid composites in wind energy applications, Materials, 2017, 10(11): 1281.

[100] S.S.P. Reddy, R. Suresh, H. M.b, B.P. Shivakumar Use of composite materials and hybrid composites in wind turbine blades, Materials Today: Proceedings, 2021, 46: 2827–2830.

[101] M. Gururaja, A.H. Rao A review on recent applications and future prospectus of hybrid composites, International Journal of Soft Computing and Engineering, 2012, 1(6): 352–355.

[102] L. Thomas, M. Ramachandra Advanced materials for wind turbine blade-A Review, Materials Today: Proceedings, 2018, 5(1, Part 3): 2635–2640.

[103] A. Corona, C.M. Markussen, M. Birkved, B. Madsen Comparative environmental sustainability assessment of bio-based fibre reinforcement materials for wind turbine blades, Wind Engineering, 2015, 39(1): 53–63.

[104] D.J. Kim, M.J. Jo, S.Y. Nam A review of polymer–nanocomposite electrolyte membranes for fuel cell application, Journal of Industrial and Engineering Chemistry, 2015, 21: 36–52.

Sanjita Wasti, Katie Copenhaver, Xianhui Zhao, Umesh Marathe,
Abdallah Ragab Barakat, Surbhi Kore, Soydan Ozcan, Uday Vaidya

2 Hybrid composite materials and their properties

2.1 Introduction

The use of fiber-reinforced polymer composites has substantially increased in recent years in different sectors such as automotive, aerospace, civil infrastructure, and sports due to their attractive properties such as high specific strength and modulus, low density, better durability, creep and fatigue resistance, and corrosion resistance [1, 2]. Fiber-reinforced polymer composites can be categorized as synthetic or natural fiber-reinforced composites. Synthetic fiber-reinforced composites exhibit excellent mechanical and thermal performance; however, they raise environmental concerns, as they are derived from fossil sources and are energy-intensive to produce. Natural fiber-reinforced composites are comparatively lighter, less expensive, and sustainable but typically have poor mechanical and thermal properties compared to synthetic fiber composites [2].

In some cases, composites with a single type of fiber reinforcement cannot attain the unique properties required for specific applications. To address this issue, composites are often hybridized by combining two or more types of fibers (synthetic–natural, natural–natural, or synthetic–synthetic) in a polymer system. Hybridization can enable tailoring of the properties of the composites by overcoming the limitations of one fiber type with the merits of the other(s). For instance, natural fiber composites

Notice: This manuscript has been authored by UT-Battelle, LLC, under contract DE-AC05-00OR22725 with the US Department of Energy (DOE). The US government retains and the publisher, by accepting the article for publication, acknowledges that the US government retains a nonexclusive, paid-up, irrevocable, worldwide license to publish or reproduce the published form of this manuscript, or allow others to do so, for US government purposes. DOE will provide public access to these results of federally sponsored research in accordance with the DOE Public Access Plan (https://www.energy.gov/doe-public-access-plan).

Sanjita Wasti, Abdallah Ragab Barakat, Tickle College of Engineering, University of Tennessee, Knoxville, Knoxville, TN, United States
Katie Copenhaver, Umesh Marathe, Soydan Ozcan, Manufacturing Science Division, Oak Ridge National Laboratory, Oak Ridge, TN, United States
Xianhui Zhao, Environmental Sciences Division, Oak Ridge National Laboratory, Oak Ridge, TN, United States
Surbhi Kore, Intel, Hillsboro, OR, United States
Uday Vaidya, Tickle College of Engineering, University of Tennessee, Knoxville, Knoxville, TN, United States; Manufacturing Science Division, Oak Ridge National Laboratory, Oak Ridge, TN, United States; Institute for Advanced Composites Manufacturing Innovation (IACMI), Knoxville, TN, United States

https://doi.org/10.1515/9783111019543-002

are often hybridized by adding glass or carbon fibers to improve mechanical, thermal, and moisture absorption properties. The thus-formed hybrid composite balances mechanical properties, weight, environmental sustainability, and cost-effectiveness.

The properties of hybrid composites are influenced by several factors such as fiber choice, properties of the matrix and fiber, manufacturing technique, fiber–matrix interface, etc. [3]. In order to achieve a desired set of properties or performance from hybrid composites, a systematic design approach is needed. Factors influencing the properties should be carefully selected while preparing the composites. This chapter provides an overview of the different types of matrices and reinforcement used for hybridization, the approaches to prepare the hybrid composites, the factors influencing hybridization, and the effect of hybridization on the properties of the final composites.

2.2 Materials

Polymer composites comprise two constituent materials: the polymer matrix and reinforcement or fillers. Hybrid composite materials are prepared by combining one or more fibers in one or more polymer matrices. These constituents must be well understood to maximize their synergy and thus the resulting composite's performance.

2.2.1 Polymer matrix

The polymer matrix is responsible for supporting the reinforcements and holding the fibers together in a solid state. It also protects the reinforcements from external environmental damage such as moisture, corrosive media, and mechanical wear. While the fibers in a reinforced composite are intended to bear most of the load, the matrix serves as a load/stress transfer medium. Polymer matrix materials can generally be classified into thermoplastics and thermosets. The polymer should be carefully selected based on the performance requirements of the end-use application, as the selected polymer system can dictate the accompanying type of reinforcement and manufacturing process to be used. Table 2.1 presents the properties of commonly used polymer matrices in fiber-reinforced composites.

2.2.1.1 Thermoplastic

Thermoplastic polymers are generally solid at room temperature and can soften and reshape upon heating without chemical reaction [4]. They offer several advantages, such as high impact resistance and toughness, damage tolerance, and ease of processing compared to thermosets [4]. Thermoplastic polymers typically have shorter manufacturing/

cycle times than thermosets and can be recycled. However, they have 500–1,000 times higher viscosity than thermoset resins, making impregnation on reinforcements difficult [4, 5]. The most commonly used thermoplastic polymers for hybrid composite systems are polypropylene (PP), polyethylene (PE), and polylactic acid (PLA).

PP is a commodity thermoplastic polymer formed by the polymerization of propylene molecules. PP is extensively used in automotive, packaging, food industry, construction, and household applications due to its outstanding chemical and moisture resistance, low density, easy processability, low processing temperature, relatively low cost, transparency, dimensional stability, and recyclability [6–10]. PE is another widely used thermoplastic polymer with several unique properties, such as high toughness, excellent chemical inertness, low coefficient of friction, ease of processing, and low electrical conductivity [11]. The properties of PE depend on its extent of branching and molecular weight, and PE is divided into low-density polyethylene (LDPE) and high-density polyethylene (HDPE) based on these characteristics [12]. LDPE has relatively long branches and irregularly packed polymer chains, making it flexible with low tensile and compressive strength and low crystallinity. HDPE consists of long, minimally branched chains that can easily pack, making it rigid with high crystalline content [11, 12]. PP and PE are regularly used as matrices for polymer composites but present a poor interface with reinforcing fibers, especially natural fibers, which is one of the significant challenges associated with these composites.

PLA is a thermoplastic aliphatic polyester, derived from the starch of agricultural plants such as corn, sugar beets, wheat, and sugarcane [13]. It is the most commonly used bioplastic and offers outstanding physical and mechanical properties, renewability and bio-degradability, processability, and high rigidity [12]. However, PLA is brittle compared to other commercially used polymers, such as PP, and has low thermal stability and low rate of crystallization [14].

2.2.1.2 Thermoset

Unlike thermoplastic polymers, thermosets cannot be softened, remelted, reshaped, or recycled upon heating once the polymerization/curing is complete, due to the covalent crosslinking between the polymer chains. Thermoset polymers exhibit superior mechanical properties (see Table 2.1), improved creep resistance, higher thermal stability, and higher chemical resistance than thermoplastics [4]. Uncured thermosets have a comparatively lower viscosity than thermoplastics, which eases resin impregnation even at low pressure. Thermoset polymers are brittle and have low toughness, leading to poor impact properties. Commonly used thermoset matrices for hybrid composites include epoxy and polyester.

Epoxy resin is one of the most commercialized classes of thermosets with unique characteristics such as small cure shrinkage, low residual stresses, and a wide temperature range of processability [15]. Additionally, epoxy resin has excellent mechanical and chemical properties, corrosion resistance, good thermal and dimensional stability, and no volatile agent release during curing compared to other thermoset systems such as polyester and phenolic [16, 17]. Epoxy resins have a longer curing time and are expensive compared to polyesters, and thus are primarily used in performance-driven industries such as aerospace [18]. Commercially, epoxy resin can be found in liquid (low viscosity) and solid (powder) form.

Unsaturated polyester resin is another commonly used thermoset matrix known for its low cost, ease of processing, capability of being cured at room temperature, availability, durability, and excellent resistance to abrasion and chemicals [19]. The major drawback of polyester resin is its significant volumetric shrinkage upon curing.

Table 2.1: Properties of different polymer matrices.

	Polymer	Density (g/cm³)	Tensile strength (MPa)	Young's modulus (GPa)	Elongation (%)	References
Thermoplastic	Polylactic acid	1.21–1.25	21–60	0.35–3.5	2.5–6	[20]
	Polyethylene	0.91–0.96	7–21	0.1–1.0	12–1200	[21]
	Polypropylene	0.90–0.92	31–45	1.1–1.95	50–145	[22]
Thermoset	Epoxy	1.1–1.6	28–130	2.7–6.0	1–6	[16, 23]
	Polyester	1.1–1.5	34–105	2.0–4.5	1–5	

2.2.2 Reinforcement

Reinforcements are the components of composite materials that act as a principal load/stress-bearing element. Reinforcing fibers can be present in different forms, such as woven or short fiber mats, continuous, discontinuous/chopped (long or short fiber), or particles [24]. The most commonly used fibers for reinforcement are carbon fiber, glass fiber, natural fibers, and basalt fibers.

2.2.2.1 Carbon fiber

Carbon fiber is extensively used to reinforce composite materials for aerospace, automotive, and sporting goods applications. Carbon fiber can be derived from different precursors such as polyacrylonitrile (PAN), pitch, and lignin; among which PAN is predominant and used commercially [25]. Based on the precursor used and the processing conditions, the properties of the fibers vary [25, 26]. For instance, PAN-based carbon fibers have high strength but comparatively lower modulus than pitch-based carbon fibers. In contrast, pitch-based carbon fibers typically have high modulus (see Table 2.2), and high thermal and electrical conductivity, but lower strength [5, 25, 26]. Due to their high cost, carbon fibers are mostly used in performance-driven areas rather than in cost-driven [5]. Furthermore, manufacturing carbon fiber is an energy-intensive process (670–704 MJ/kg) that results in significant carbon emissions, which further contribute to greenhouse gas (GHG) emissions [27].

2.2.2.2 Glass fiber

Glass fiber is a synthetic amorphous fiber used in a wide range of applications such as automotive, marine, wind energy, and aerospace. Glass fiber can be classified into different types such as E-glass, S-glass, C-glass, D-glass, etc., based on its composition [28]. Each of these types has distinct properties that make them suitable for specific applications. For example, E-glass has high strength and electrical resistivity and is commonly used for general-purpose applications, including electrical insulation [28]. S-glass fiber is a commonly used glass fiber that has comparatively higher strength, stiffness, and corrosion resistance and is more expensive (5–7 USD/kg) than E-glass fiber (0.75–1.2 USD/kg), restricting its use primarily to aerospace applications [5, 28, 29]. Despite its good mechanical properties, glass fiber has a comparatively higher density than carbon fiber and natural fibers, which limits its application to those where lightweighting is crucial. Additionally, glass fiber is more moisture-sensitive than carbon fiber, which affects the strength of the fiber and the interface between the fiber and the matrix [5].

2.2.2.3 Basalt fiber

Basalt fiber is a high-performance inorganic silicate fiber derived from natural volcanic basalt rock without additional components [5, 29]. Basalt fiber has mechanical properties similar to glass fiber (see Table 2.2) but differs in atomic composition [5, 30]. In addition to mechanical properties, basalt fiber also offers excellent heat, chemical, and fatigue resistance, making it a promising reinforcing alternative to glass fiber for several applications such as marine, automotive, and sporting [30]. However, compared to common glass fibers (E-glass, 0.75–1.2 USD/kg), basalt fiber (2.5–3.5 USD/kg) is difficult to spin, which makes it more expensive [5, 29]. Also, it has a slightly higher density (2,800 kg m^{-3}) than glass fiber (2,400–2,700 kg m^{-3}).

2.2.2.4 Natural fiber

In recent years, natural fibers have been gaining popularity in composite manufacturing due to the increasing environmental concerns associated with the production of carbon and glass. Based on the origin/source, natural fiber can be divided into three categories: plant, animal, and mineral fiber. Among these, plant fibers are the most commonly used natural fibers due to their relatively low cost and high mechanical properties compared to animal or mineral fibers [14]. Plant fiber is further categorized into sub-categories, such as bast, leaf, fruit, and seed, based on the part of the plant from which it is extracted. The properties of commonly used plant fibers are presented in Table 2.2. Natural fibers possess unique properties such as low density, biodegradability, natural abundance, low cost, and good thermal and sound insulation. However, the enormous range of natural fiber compositions, microstructures, geometries, and variations in performance within the same fiber types make their selection, optimization, and implementation into large-scale composite processing complex [31]. Additionally, their poor compatibility with polymer matrices, low mechanical properties, low thermal stability, and high moisture-absorbing tendency, compared to synthetic fibers, limits their range of application [4].

Table 2.2: Properties of different fibers used for reinforcement.

Fiber	Type	Density (g/cm^3)	Tensile strength (MPa)	Young's modulus (GPa)	Elongation (%)	Fiber diameter (μm)	References
Carbon fibers	PAN	1.73–2.00	2,700–7,100	200–700	0.7–2.2	5–7	[5, 25]
	Pitch-based	1.57–2.20	1,000–4,000	400–940	0.3–3	5–12	
Glass fibers	E-Glass	2.54–2.58	1400–3,500	72–79	1.8–4.8	6–21	[5, 28, 29, 32]
	S-Glass	2.5	4,570	86	2.8	6–21	
Natural fibers	Coir	1.2–1.5	95–240	2.8–6	15–51	150–250	[12, 33–36]
	Bamboo	0.6–1.4	140–800	11–36	2.5–3.7	240–330	
	Jute	1.23–1.50	320–860	8–78	1–2.5	40–350	
	Flax	1.4–1.5	343–2,000	27.6–103	1.2–3.3	10–50	
	Hemp	1.4–1.5	270–1,100	23.5–90	2–4	17–23	
Basalt fibers		2.63–3.00	2,800–4,840	79–110	3.1	6–21	[5, 29]

2.3 Hybridization approaches

Hybrid composites can be prepared using a wide range of composite manufacturing techniques. The manufacturing method selection for hybrid composites depends on factors such as the polymer matrix (thermoplastics or thermosets), reinforcement/fiber type (synthetic–synthetic, natural–natural, or natural–synthetic), reinforcement forms (woven fabric, continuous, long, short, and particles), dimension of the required composite part, production rate, and cost [37]. The implemented manufacturing technique greatly influences the final composite properties. Hand layup, followed by compression molding and vacuum-assisted resin infusion with woven or nonwoven fiber mats, is the most common manufacturing technique for preparing thermoset-based hybrid fiber composites. In contrast, thermoplastic hybrid composites are typically manufactured by compounding (using a batch mixer or twin screw extruder) polymer pellets with chopped fiber or particles, followed by compression or injection molding. Table 2.3 lists several manufacturing techniques for preparing synthetic and natural fiber hybrid composites. Details on hybridization approaches using different manufacturing techniques can be found in various chapters in this book.

Table 2.3: Processing/manufacturing techniques used for making hybrid composites.

Fibers	Polymer	Reinforcement form	Manufacturing process	Reference
Banana/jute	Epoxy	Unidirectional continuous fiber	Hand layup, compression molding	[38]
Palmyra palm leaf stalk fiber/jute	Polyester	Unidirectional continuous fiber	Hand layup, compression molding	[39]
Kenaf/pineapple leaf fiber (PALF)	PP	Short fiber	Compounding, compression molding	[40]
Kenaf/coir	PP	Short fiber	Compounding, compression molding	[41]
Jute/glass	Epoxy	Woven fabric	Hand layup	[42]
Banana/sisal/glass	Epoxy	Short fiber and woven roving	Compression molding	[43]
Sisal/jute/glass	Polyester	Long fiber and unidirectional mat	Hand layup, compression molding	[44, 45]
Flax/glass, jute/glass	Epoxy	Woven fabrics	Vacuum-assisted resin infusion	[46]
Pine/agave	HDPE	Particles	Twin screw extrusion, injection molding	[47]
Banana/sisal	PLA	Short fibers	Twin screw extrusion, injection molding	[48]
Bamboo/carbon	PP	Short fibers and continuous tape	Extrusion compression overmolding	[49]
Bamboo/carbon	PP	Long fibers	Wet laying, compression molding	[50]

2.4 Factors influencing hybridization/compatibility

Barriers to improved performance in hybrid fiber-reinforced composites are similar to those in composites reinforced with a single type of fiber. The performance of hybrid fiber composites depends on the properties of each fiber type present, their volume ratios, their dispersion within the polymer matrix, their aspect ratios, the degree of intermingling of the different types of fibers, the fiber orientation, and the quality of the interface between the fibers and the matrix [3].

The interfacial region between the polymer matrices and the reinforcing fibers is commonly referred to as an interphase in which the chemical composition and microstructure differ from that of the bulk polymer matrix, particularly when surface treatment or sizing is applied to the fibers [51]. The properties of the interphase determine how efficiently load can be transferred from the matrix to reinforcing fibers and the extent of stress concentration, strongly influencing the mechanical properties of the composite [51, 52]. Additionally, the interphase region dictates the thermal stability of the composite and its propensity to absorb moisture. It experiences different thermal expansion across the interphase, depending on the properties of the fibers, polymer matrix, and any fiber coatings or additives [17, 52].

Understanding and optimizing the properties of the interphase region such as the degree and strength of adhesion at the polymer/fiber interface is crucial in the design of all fiber-reinforced polymer composites. The interfacial adhesion between the fibers and the polymer matrices is often classified into four types: interdiffusion, electrostatic adhesion, chemical adhesion, and mechanical interlocking. Adhesion between a polymer matrix and the fiber surface within the interphase region can be through any of these methods or combinations thereof [52]. Interdiffusion is a physical adhesion mechanism that relies on the wettability of the polymer on fiber surfaces, which is governed by the polar (related to acid–base interactions) and dispersive (related to Van der Waals interactions) components of the surface energy of both the polymer matrix and the reinforcing fibers. Chemical modifications or additives can produce composites to increase the wettability of the fibers by the polymer matrices and promote adhesion through Van der Waals attraction [17, 52]. Electrostatic adhesion occurs between positively and negatively charged moieties but has yet to be widely studied in fiber-reinforced composites to date. Chemical adhesion refers to covalent bonding between the fiber surface groups and the surrounding polymer matrix. This can be achieved through chemical treatments to introduce bonding sites to the fibers or matrix or the addition of coupling agents [52]. Finally, mechanical interlocking can occur between a polymer and the rough fiber surface in which the polymer penetrates cracks or holes on the fiber, anchoring the matrix to the fiber surface. Chemical and physical treatments are often applied to fibers to roughen their surfaces and promote adhesion through mechanical interlocking [17, 52]. These treatments can often serve multiple purposes, such as chemical treatments that remove hydrophilic or highly polar moieties from a fiber surface, while simultaneously roughening it. The most commonly reported chemical treatments in all types of fiber-reinforced composites include the use of silanes, alkaline substances, and peroxides, while common physical treatments include the applications of plasma or corona, calendaring, heat application, and high energy electron, laser, or ultraviolet irradiation. Physical treatments can provide the advantage of altering the surface structure of the fibers, while preserving their composition [17, 37, 52].

2.4.1 Common treatments for synthetic fibers

Synthetic fibers such as carbon and glass suffer from poor adhesion with polymer matrices due to their smooth, inert surfaces. As such, both fibers are typically coated in a sizing agent during their production, and numerous physical treatments have been shown to roughen their surfaces for better adhesion with polymer matrices [53]. Sizing agents serve not only to provide chemical compatibility between the fiber surfaces and the polymer matrices but also to protect fibers during handling and processing to preserve their length and inherent strength [54]. The surface of carbon fibers comprises highly crystallized graphitic planes, presenting a chemically inert and low-energy surface [55, 56]. Chemical surface treatments applied to carbon fibers for use in composites include both sizing agents and acid treatments, the latter of which can serve to roughen and oxidize surfaces or initiate polymer grafting onto surfaces in the presence of other polymerization agents. Sizing agent formulations are often tailored for polymer matrices such as epoxies or polyamides. However, the most common sizing agents used with carbon fibers are epoxy-, polyamide-, or phenoxy-based and consist of emulsions or dispersions of reactive agents and/or oligomers. In some cases where carbon fiber is used in a very high-temperature process, such as melt processing of polyetheretherketone (PEEK) at 400 °C, more thermally stable sizing agents must be applied [53]. Oxidation through acid treatment or other chemical applications increases the surface activity and functional sites of carbon fibers, but it can also easily damage the fibers if not carefully controlled. In addition to acids such as nitric or sulfuric acid, carbon fibers can also be oxidized using peroxides, hydroxides, and electrochemical treatments, among others [51, 56]. Numerous other treatments, including the use of plasma, heat treatments, high energy irradiation, and deposition of multi-scale surface structures through chemical vapor deposition, electrophoretic deposition, or electrospraying have been shown to roughen and further activate the surfaces of carbon fibers, leading to increases in wettability, surface energy, and mechanical interlocking, ultimately improving stress transfer and overall performance of the composite [51, 55, 56].

Almost all glass fibers used as composite reinforcements are silica-based and, as such, are amorphous with nearly isotropic properties. Sizing agent formulations for glass fibers are typically aqueous dispersion or emulsions and contain a film-forming moiety and a coupling agent [53]. Formulations can also include lubricants and agents for providing antistatic, wetting, and roughening properties [57]. Similar to carbon fiber sizing agents, the components of glass fiber sizings are selected with respect to the target polymer matrix. The film-forming component of glass fiber sizing is typically a polymeric compound. Epoxy-based film formers are the common agents used with glass fibers in thermoset matrices based on epoxy, polyester, and vinyl ester compounds. In contrast, polyvinyl alcohol (PVA)- or polyurethane (PU)-based film-forming agents are most commonly used in glass fiber sizings for thermoplastic composites such as polyamides or polyolefins [53].

2.4.2 Common treatments for natural fibers

Most natural fibers have numerous commonalities (such as poor fiber compatibility with matrix, low mechanical strength compared to synthetic fibers, lower thermal stability, and high moisture absorption) that impede their more widespread use in polymer composites and can necessitate their hybridization with synthetic fibers. Most natural fibers are hydrophilic and hygroscopic, owing to the abundant hydroxyl groups on their surfaces. The high polarity and hydrophilicity of natural fiber surfaces often lead to their poor dispersion and low interfacial adhesion within typical nonpolar, hydrophobic polymer matrices [31, 37, 58].

Chemical treatments or the addition of coupling agents are often deemed necessary to reduce the water uptake of natural fibers and improve their interfacial interactions with polymer matrices. Alkaline treatments, also referred to as mercerization, are by far the most common treatment employed for natural fibers, and sodium hydroxide (NaOH) is the most frequently reported alkaline agent in literature. Alkaline treatment roughens the fiber surfaces and removes noncellulosic substances such as lignin, pectin, and hemicellulose. Removal of these groups and other impurities can expose more active surface groups, improving the polymer and fibers' chemical compatibility, and enhancing the wetting of the resin on fiber surfaces. Similar to carbon and glass fibers, roughening of natural fiber surfaces can promote interfacial adhesion via mechanical interlocking [31, 37]. Fiber sizings, coupling agents, and compatibilizers are also used with natural fibers to improve their interactions with polymer matrices and enhance the performance of the resulting composites. A wide array of sizing and compatibilization formulations has been used in natural fiber-reinforced polymer composites in literature, the most common of which are silanization and acetylation. Both treatments introduce functional groups to fiber surfaces to improve adhesion with polymer matrices. Acetylation treatments can additionally remove some lignin and hemicelluloses from natural fiber surfaces. Finally, maleic anhydride-based coupling agents are often used in natural fiber-reinforced composites and can be easily tailored to the polymer matrix of interest [31, 52, 59].

2.5 Properties of hybrid composites

2.5.1 Mechanical properties

Various natural fiber types have been hybridized with natural or synthetic fibers to reinforce polymer composites for different applications. The selected natural fiber species strongly influences the resulting mechanical performance of the hybrid composites, and the combination of some types of natural fibers has been shown to have a synergistic effect [60]. For example, Devireddy and Biswas studied hybrid composites consisting of banana and jute fibers with epoxy as the matrix. The hybrid composites had a higher tensile strength than the banana/epoxy composite and jute/epoxy composite alone [38]. A similar synergistic effect was observed by Arthanarieswaran et al. between sisal and banana fibers in an epoxy matrix [43]. The tensile strength of the sisal/banana/epoxy hybrid composites was comparatively higher (25 MPa) than banana/epoxy (21 MPa) and sisal/epoxy (23 MPa). In another study, Shanmugam et al. used Palmyra palm leaf stalk fiber (PPLSF) and jute fiber to reinforce polyester [39]. The jute-to-PPLSF mass ratio varied from 0:100 to 25:75, 50:50, 75:25, and 100:0. It was shown that increasing the jute fiber content (from 0:100 to up to 75:25) led to an increase in the tensile strength of the composites from 57 to 83 MPa. The jute fiber-reinforced polyester composite had a tensile strength of 77 MPa, lower than the optimum hybrid PPLSF–jute composite [39]. However, the reason for these improvements in strength after hybridization in each of these studies has not been fully elucidated. Natural fibers with poor mechanical strength are often hybridized with another natural fiber(s) with high mechanical strength to achieve the required mechanical performance, while maintaining environmental sustainability and cost. For instance, coir fiber has comparatively lower strength than other natural fibers but is a byproduct/waste product from the high-volume, nonseasonal coconut crop. Coir fiber has high elongation at break, high weather resistance, and is less expensive than other natural fibers. The strength of coir-reinforced composite is typically insufficient for advanced applications, and it is often hybridized with other high-performing but more expensive natural fibers. For example, Islam et al. added kenaf fiber, a natural fiber with superior mechanical properties than coir fiber, to improve the performance of PP-based composites [41].

The mechanical properties of hybrid natural fiber composites are greatly influenced by fiber pretreatment. In a study by Asaithambi et al. [48] authors observed that the tensile strength and modulus of banana/sisal/PLA hybrid composites increased from 57 MPa to 79 MPa and from 1.7 GPa to 4.1 GPa, respectively, on treating banana/sisal fibers with alkali followed by benzoyl peroxide. This was attributed to an increased wettability and compatibility with the matrix after the chemical treatment. Similarly, Saw et al. observed improvement in the tensile and flexural strength of jute/coir/epoxy hybrid composites by 78% and 61%, respectively (compared to untreated hybrid fiber composites), after alkali treatment of the jute fiber and chemical modification of coir fiber with furfuryl alcohol [61]. The tensile and flexural strength improvements were attributed to an improved fiber–matrix interface. In contrast, the impact strength was higher for the untreated hybrid fiber-reinforced composites compared to the treated one, which could be attributable to an increased energy absorption by the weaker untreated fiber–matrix interface [61].

Fiber weight or volume content is one factor affecting the extent of improvement in the mechanical properties of hybrid composites. For instance, Sathishkumar et al. found insignificant improvement in the tensile strength of snake grass fiber-reinforced polyester composites upon hybridization with coir fiber and banana fiber at a 10% volume fraction [62]. However, upon increasing the fiber volume fraction to 20% and 25%, the authors observed maximum improvement in tensile strength of snake grass/ banana fiber composites and snake grass/ coir fiber composites, respectively [62]. Similarly, Devireddy et al. observed an optimal tensile and flexural strength at 30 wt% fiber loading (compared to 10, 20, and 40 wt%) and decreased on further increasing the fiber loading, attributable to fiber entanglement and agglomeration that hampered stress transfer between the fiber and the matrix [38]. Additionally, variation in the ratios of different fibers used for hybridization at a constant fiber volume/weight fraction also impacts the properties of their associated composites [38]. For example, in a study, Saw et al. produced a series of epoxy composites reinforced with 30 vol% jute and coir with jute: coir volume fractions at 100:0, 80:20, 65:35, 50:50, 35:65, and 0:100. The authors reported significant improvements in the tensile and flexural properties of hybrid composites at equal volume fractions of jute and coir fibers [61].

To meet performance requirements for advanced applications, natural fiber composites are often hybridized with synthetic fibers, as adding a small amount of synthetic fiber can significantly enhance their mechanical properties. For instance, Wasti et al. observed ~ 152% and ~ 164% enhancement in flexural strength and modulus, respectively, upon overmolding two layers of unidirectional textile grade carbon fiber (TCF)/ PP tape on bamboo fiber (20 wt%)/PP composites [49]. This was attributed to the higher resistance unidirectional TCF offers upon compressive loads. The average content of TCF in the hybrid composite was approximately 2 wt%. Similarly, upon adding just 5 wt % of glass fiber, AlMaadeed et al. observed a 16% improvement in tensile strength of wood flour/recycled PP composites [63]. SEM images of fractured surfaces suggested that a higher aspect ratio of the glass fiber and a better interface between the glass fiber and the PP matrix were responsible for increased strength upon hybridization. Interfacial bonding between the fibers and the matrix must be optimized to fully exploit the potential of reinforcing agents in the hybrid composites. Valente et al. found a decrease in the flexural strength of recycled glass/wood flour-reinforced PP and LDPE composites on increasing the wood flour content in the hybrid system [64]. The authors found that the poor interface between the wood flour and the polymer matrix made ineffective transfer of stress from the matrix to the fiber, and resulted in decreased flexural strength. A similar observation was made by Ghasemzadeh-Barvarz et al. with glass/flax/PP composites, where an increase in the ratio of glass/flax or increasing content of glass fiber had a negligible effect on the tensile strength of hybrid composite due to the poor interface between PP and the glass fiber [65].

Fiber length is another important factor that influences the mechanical properties of hybrid composites. Jarukumjorn et al. found an insignificant increase in tensile and flexural properties of sisal/glass/PP hybrid composites on increasing the glass fiber content [66]. The authors theorized the lack of improvement was attributable to breakage of the glass fiber during the composite manufacturing process (compounding followed by injection molding). In another study, Kahl et al. observed enhancements in the tensile strength and modulus, upon hybridization of regenerated cellulose/PP with glass fibers [67]. However, increasing the content of glass fiber in the hybrid system did not lead to enhancement in strength, which might be due to glass fiber attrition during composite manufacturing (twin screw extrusion-injection molding). In addition, the authors observed an increase in the elongation-at-break and Charpy impact properties at higher cellulose content in the hybrid system, attributable to the longer fiber length and higher elongation at the break of the cellulose fibers [67].

The fiber orientation and stacking sequence of laminates have also been reported to affect the mechanical properties of hybrid fiber composites. For example, Ramesh et al. found that hybrid composites with fibers oriented at 0° exhibited better tensile, flexural, and impact properties compared to composites with fibers oriented at 90° [44]. Similarly, Ramnath et al. found that the mechanical properties of intralayer abaca–jute–glass fiber can be optimized by altering the fiber orientations [68]. The authors found that hybrid composites with 45° orientation (in the second layer) had higher tensile and flexural properties, followed by composites with parallel and perpendicular fiber orientation. Studies have suggested that the mechanical properties and failure mechanism of hybrid composites can be modified and tuned by carefully designing the stacking sequence of fiber laminates [43, 46, 69]. Pinto et al. found that for carbon/hemp hybrid composites, the flexural modulus of the hybrid composite can be improved by placing the hemp layer close to the neutral axis, whereas improvement in absorbed energy during flexural and impact dynamic conditions can be obtained by placing the hemp layers on the top side of the laminate [69]. In another study by Selver et al., the authors did not observe any difference in tensile strength and modulus of flax/glass and jute/glass hybrid composites at different stacking sequences [46]. However, the flexural strength was higher for the composites with glass fiber on the outer layers (glass/flax/glass or glass/jute/glass) than on the middle layer (flax/glass/flax or jute/glass/jute). This might be due to the higher bending stiffness of glass fiber compared to natural fibers. In flexural testing, failure mostly starts at the outer layer. It propagates to the inner layer, so arranging the outer layers with material with high bending stiffness yields higher flexural strength [46].

2.5.2 Thermal properties

The thermal properties of interest in polymer composites include degradation, melting, crystallization, and deflection behaviors, when subjected to a range of temperatures. Several researchers have studied the influence of fiber hybridization on the thermal properties of the composites and have shown that fiber type, fiber content, matrix type, and fiber surface treatment conditions significantly affect the hybrid composites' thermal stability and degradation behavior. Plant fibers such as flax, hemp, banana, etc., undergo multistage decomposition due to moisture and lignocellulosic components such as hemicellulose, cellulose, and lignin. Therefore, weight loss occurs in multiple stages for hybrid composites comprising natural fibers. For instance, a two-step weight loss was observed in a study by Neto et al. on jute/ramie/epoxy, Jute/curaua/epoxy, and jute/sisal/epoxy composites [31]. An initial weight loss ranging from 30–180 °C due to moisture present on fiber components was followed by a second, ranging from 215–450 °C, due to the degradation of hemicellulose, cellulose, and lignin. Researchers have also studied the effect of chemical treatments (such as alkali, alkali + silane, and benzoylation) of natural fibers on the thermal stability of hybrid natural fiber-reinforced composites. They found that chemical treatment can remove less thermally stabile constituents from the fiber surface, enhancing the thermal stability of the fibers and their composites [31, 70].

One of the limitations of natural fibers in composites is their lower thermal stability. Natural fibers are often hybridized with synthetic fibers to improve their thermal and mechanical performance. Braga et al. prepared epoxy/jute/glass composites at ratios of 69:31:0, 68:25:7 and 64:18:19 and observed reductions in weight loss at 100 °C (from 1.95% to 1.52% and 1.27%, respectively), 200 °C (from 9% to 7.76% and 6.57%, respectively), and 450 °C (from 70.7% to 68.97% and 63.54%, respectively). The final residue was increased from 6.48% to 24.19%, on increasing the glass fiber content from 0 wt% to 19 wt% [42]. Similarly, Atiqah et al. reported a shift in the onset degradation temperature to higher temperatures and an increase in char residue, with the addition of glass fiber to sugar palm-reinforced TPU composites [71].

2.5.3 Moisture absorption properties

Water absorption in polymer composites occurs due to multiple phenomena, such as water absorption by fiber strands, capillary transport of water into the flaws and gaps of the polymer-matrix interface, and diffusion of water molecules into the polymer chain [65].

In the case of natural fiber-reinforced composites, natural fibers absorb atmospheric moisture due to their hydrophilic nature, leading to swelling and further deterioration of the polymer/fiber interface. Fiber swelling disrupts stress transfer from the matrix and can lead to microcracks within the composite [17, 37, 58]. Additionally, water present within a natural fiber or at its surface can lead to leeching of its water-soluble components, exacerbating debonding of the matrix from the fiber and further deteriorating the composite properties [17, 58]. For natural fiber hybrid composites, studies have suggested that the addition and increase in the concentration of natural fiber typically increases the water uptake of the composites. In a study by Islam et al., the authors noted higher water absorption for hybrid fiber-reinforced composites (Kenaf/coir/PP) compared to single fiber-reinforced composites (kenaf/PP and coir/PP), which might be due to the mixing of two hydrophilic fibers in the composites [41].

Additionally, researchers have found that along with the fiber content, factors such as the composition of natural fibers (cellulose, hemicellulose, and lignin content) and fiber treatments also influence the water absorption behavior of their composites. Feng et al. studied the moisture uptake behavior of kenaf and pineapple leaf fiber (PALF)-reinforced polypropylene composites (total fiber content of 30 wt%) and observed an increase in moisture uptake with increase in PALF content in the hybrid composites. The authors correlated this phenomenon with the higher cellulose and lower lignin content of the PALF fibers compared to kenaf. Hydroxyl groups in cellulose form hydrogen bonds with water, increasing water uptake of overall composites. At the same time, lignin is a more hydrophobic moiety due to its abundance of phenolic groups [40]. Furthermore, Senthil Kumar et al. studied the effect of alkaline treatment on the water absorption capacity of hybrid banana and sisal fiber-reinforced composites in a cashew nut cell oil resin, where they found a decrease in water absorbability with fiber treatment [72]. This was attributed to the improved interface between the fiber and matrix, upon alkali treatment.

Natural fibers-reinforced composites are also often hybridized with synthetic fibers to reduce the water uptake behavior of the composites. Synthetic fibers are hydrophobic in nature and have negligible water absorption capacity. In these hybrid composite systems, synthetic fibers can act as a barrier, preventing direct contact between the natural fibers and water molecules. Ghasemzadeh- Barvarz et al. [65] and Braga et al. [42] observed reduced water absorption tendency of PP/flax and epoxy/jute composites, upon hybridizing with glass fibers. Ramesh et al. observed a similar gradual decrease in water uptake behavior of epoxy/banana fiber composites upon substituting 20–80 wt % of banana fiber with carbon fiber [73]. Furthermore, studies have also suggested that the moisture absorbing tendency of hybrid fiber composites is affected by fiber configuration or layup [1, 74]. For example, for the same glass and flax fiber content, Naga Kumar et al. observed that composites with flax fiber fabrics on the outer layer exhibited higher water uptake capacity than the composites with glass fibers [74].

2.5.4 Other properties

Noise and vibration are the other important properties of materials and can be exploited to fabricate soundproof internal parts for vehicles. The dampening performance of polymers can be extended by combining them with hybrid reinforcing materials. Hybrid polymer composites with natural and synthetic fiber reinforcements and micro- and nano-sized fillers offer a versatile solution for applications demanding superior vibration damping and acoustical properties. These composites provide structural robustness, and excel in reducing vibrations and noise, making them invaluable in industries where noise control, structural integrity, and weight savings are essential considerations. Their adaptability and ability to tailor properties make them an exciting materials science research and development area.

Selver et al. fabricated composite epoxy base laminate using vacuum-assisted resin transfer molding with plain weave of glass fibers, plain weave of jute, and canvas weave of flax fibers. Moreover, the different stacking sequences were explored to understand their effect on sound absorption coefficient and transmission rate. It was observed that natural fiber and hybrid composites showed higher transmission losses compared to synthetic counterpart. It was reported that in the case of stacking sequence, keeping natural fiber outside bestows better sound insulation [75]. Using the solution-casting method, Kim et al. developed PP-based hybrid nanocomposite with MWCNTs (0.1, 0.5, and 0.7 wt%) and Closite clay (0.9, 4.8, and 6.5 wt%). Sound transmission loss for nanocomposite was 4.8 wt%; closite, 0.5 wt%; and MWCNT was about 15–21 dB higher than pristine PP. Moreover, the soundproofing properties achieved in the study could be attributed to the efficient dispersion and distribution of nanomaterials throughout the PP [76]. In another study, Kim et al. fabricated PP/MWCNT/exfoliated graphite and characterized electromagnetic shielding and soundproofing. It was found that composite with 10 wt% MWCNT and 10 wt% of exfoliated graphite showed 5 dB higher sound transmission loss than pure PP [77]. Haris et al. studied the hybrid composite of PP/carbon fiber (plain weave)/flax fiber (plain weave) [78]. Two stacking iterations were explored, i.e., five stacks of flax and PP and three stack sandwiches of flax-PP and face of carbon fiber/polypropylene. Hybrid composites showed better overall sound transmission loss compared to just flax composites over the spectrum of frequencies (50–1,600 Hz) [78].

2.6 Conclusion

This book chapter reviews hybrid materials, focusing on the broad spectrum of reinforcements, polymers, hybridization approaches, surface treatments of reinforcements, their subsequent effect on mechanical, thermal, and water absorption properties, and noise and acoustic properties. Challenging applications, demand of properties, drives toward sustainability, circular economy, and environmental concerns are the significant driving forces to develop hybrid polymer composites. Hybrid combinations of fillers and fibers tend to show synergistic performance properties such as mechanical, thermal, and moisture absorption. These performance properties depend on hybridization approach, manufacturing technologies involved, compatibility, different types and sizes of reinforcing fibers/particles and pre-applied surface treatments like acid treatment, alkali treatment, and sizing. Mechanical properties are the major break-even points for any combination of hybrid reinforcement. The mechanical properties of composites with a combination of different natural fibers vary. However, it can be improved more in combination with synthetic fibers. For thermal properties, the addition of multi-type fibers, synthetic and natural, leads to multistaged degradation. Moreover, their surface treatment bestows improved thermal stability by removing thermally vulnerable components such as lignin, in the case of natural fibers. It was found that synthetic fibers are added to natural fiber composites to improve their water absorption properties. Moreover, pretreatment on the natural fibers, composition of natural fibers, and their combination (or content) with synthetic fibers affect the water absorption behavior of the hybrid composites.

Acknowledgment: The authors acknowledge the support from the US Department of Energy (DOE), Office of Energy Efficiency and Renewable Energy, and Advanced Materials and Manufacturing Office. This manuscript has been authored by UT-Battelle, LLC, under contract DE-AC05-00OR22725 with the US Department of Energy (DOE). The US government retains and the publisher, by accepting the article for publication, acknowledges that the US government retains a nonexclusive, paid-up, irrevocable, worldwide license to publish or reproduce the published form of this manuscript, or allow others to do so, for US government purposes. DOE will provide public access to these results of federally sponsored research in accordance with the DOE Public Access Plan (http://energy.gov/downloads/doe-public-access-plan).

References

[1]   M.K. Gupta, M. Ramesh, S. Thomas Effect of hybridization on properties of natural and synthetic fiber-reinforced polymer composites (2001–2020): A review, Polymer Composites, 2021, 42: 4981–5010. https://doi.org/10.1002/pc.26244.

[2]   S.O. Ismail, E. Akpan, H.N. Dhakal Review on natural plant fibres and their hybrid composites for structural applications: Recent trends and future perspectives, Composites Part C: Open Access, 2022, 9: 100322. https://doi.org/10.1016/j.jcomc.2022.100322.

[3]   M. Jawaid, H.P.S. Abdul Khalil Cellulosic/synthetic fibre reinforced polymer hybrid composites: A review, Carbohydrate Polymers, 2011, 86: 1–18. https://doi.org/10.1016/j.carbpol.2011.04.043.

[4]   A. Gholampour, T. Ozbakkaloglu A Review of Natural Fiber Composites: Properties, Modification and Processing Techniques, Characterization, Applications, Springer, US, 2020.

[5]   Y. Swolfs, I. Verpoest, L. Gorbatikh Recent advances in fibre-hybrid composites: Materials selection, opportunities and applications, International Materials Reviews, 2019, 64: 181–215. https://doi.org/10.1080/09506608.2018.1467365.

[6]   A.L. Andrady, M.A. Neal Applications and societal benefits of plastics, Philosophical Transactions of the Royal Society of London. Series B, Biological Sciences, 2009, 364: 1977–1984. https://doi.org/10.1098/rstb.2008.0304.

[7]   Q.T.H. Shubhra, A.K.M.M. Alam, M.A. Quaiyyum Mechanical properties of polypropylene composites: A review, Journal of Thermoplastic Composite Materials, 2013, 26: 362–391. https://doi.org/10.1177/0892705711428659.

[8]   A. Chatterjee, S. Kumar, H. Singh Tensile strength and thermal behavior of jute fibre reinforced polypropylene laminate composite, Composites Communications, 2020, 22: 100483. https://doi.org/10.1016/j.coco.2020.100483.

[9]   J. Xiao, Y. Chen New micro-structure designs of a polypropylene (PP) composite with improved impact property, Material Letters, 2015, 152: 210–212. https://doi.org/10.1016/j.matlet.2015.03.101.

[10]   E. Watt, M.A. Abdelwahab, M.R. Snowdon et al Hybrid biocomposites from polypropylene, sustainable biocarbon and graphene nanoplatelets, Scientific Reports, 2020, 10: 1–13. https://doi.org/10.1038/s41598-020-66855-4.

[11]   P.N. Khanam, M.A.A. AlMaadeed Processing and characterization of polyethylene-based composites, Advanced Manufacturing: Polymer & Composites Science, 2015, 1: 63–79. https://doi.org/10.1179/2055035915Y.0000000002.

[12]   M. Li, Y. Pu, V.M. Thomas et al Recent advancements of plant-based natural fiber–reinforced composites and their applications, Composites Part B: Engineering, 2020, 200: 108254. https://doi.org/10.1016/j.compositesb.2020.108254.

[13]   S. Wasti, S. Adhikari Use of Biomaterials for 3D Printing by Fused Deposition Modeling Technique: A Review, Frontiers in Chemistry, 2020, 8: 1–14. https://doi.org/10.3389/fchem.2020.00315.

[14]   J.S.S. Neto, H.F.M. De Queiroz, R.A.A. Aguiar, M.D. Banea A review on the thermal characterisation of natural and hybrid fiber composites, Polymers (Basel), 2021, 13, https://doi.org/10.3390/polym13244425.

[15]   H. Sukanto, W.W. Raharjo, D. Ariawan et al Epoxy resins thermosetting for mechanical engineering, Open Engineering, 2021, 11: 797–814. https://doi.org/10.1515/eng-2021-0078.

[16]   T.G. Yashas Gowda, M.R. Sanjay, K. Subrahmanya Bhat et al Polymer matrix-natural fiber composites: An overview, Cogent Engineering, 2018, 5, https://doi.org/10.1080/23311916.2018.1446667.

[17]   J. Neto, H. Queiroz, R. Aguiar et al A review of recent advances in hybrid natural fiber reinforced polymer composites, Journal of Renewable Materials, 2022, 10: 561–589. https://doi.org/10.32604/jrm.2022.017434.

[18]   N. Ramadan, M. Taha, A.D. La Rosa, A. Elsabbagh Towards selection charts for epoxy resin, unsaturated polyester resin and their fibre-fabric composites with flame retardants, Materials (Basel), 2021, 14: 1–44. https://doi.org/10.3390/ma14051181.

[19]   P. Pączkowski, A. Puszka, B. Gawdzik Effect of eco-friendly peanut shell powder on the chemical resistance, physical, thermal, and thermomechanical properties of unsaturated polyester resin composites, Polymers (Basel), 2021, 13, https://doi.org/10.3390/polym13213690.

[20]   S. De, B. James, J. Ji et al Biomass-derived Composites for Various Applications, 1st ed, Elsevier Inc, 2023.

[21]   S.L. Favaro, A.G.B. Pereira, J.R. Fernandes et al Outstanding impact resistance of post-consumer HDPE/multilayer packaging composites, Materials Sciences and Applications, 2017, 08: 15–25. https://doi.org/10.4236/msa.2017.81002.

[22]   K.M.F. Hasan, P.G. Horváth, T. Alpár Potential natural fiber polymeric nanobiocomposites: A review, Polymers (Basel), 2020, 12, https://doi.org/10.3390/POLYM12051072.

[23]   B. Ravishankar, S.K. Nayak, M.A. Kader Hybrid composites for automotive applications – A review, Journal of Reinforced Plastics and Composites, 2019, 38: 835–845. https://doi.org/10.1177/0731684419849708.

[24]   D.O. Bichang'A, F.O. Aramide, I.O. Oladele, O.O. Alabi A review on the parameters affecting the mechanical, physical, and thermal properties of natural/synthetic fibre hybrid reinforced polymer

composites, Advance Material Science and Engineering, 2022, 2022, https://doi.org/10.1155/2022/7024099.

[25]   K. Naito, Y. Tanaka, J.M. Yang, Y. Kagawa Tensile properties of ultrahigh strength PAN-based, ultrahigh modulus pitch-based and high ductility pitch-based carbon fibers, Carbon N Y, 2008, 46: 189–195. https://doi.org/10.1016/j.carbon.2007.11.001.

[26]   X. Qin, Y. Lu, H. Xiao et al A comparison of the effect of graphitization on microstructures and properties of polyacrylonitrile and mesophase pitch-based carbon fibers, Carbon N Y, 2012, 50: 4459–4469. https://doi.org/10.1016/j.carbon.2012.05.024.

[27]   S. Das Life cycle assessment of carbon fiber-reinforced polymer composites, International Journal of Life Cycle Assessment, 2011, 16: 268–282. https://doi.org/10.1007/s11367-011-0264-z.

[28]   T.P. Sathishkumar, S. Satheeshkumar, J. Naveen Glass fiber-reinforced polymer composites – A review, Journal of Reinforced Plastics and Composites, 2014, 33: 1258–1275. https://doi.org/10.1177/0731684414530790.

[29]   H. Liu, Y. Yu, Y. Liu et al A review on basalt fiber composites and their applications in clean energy sector and power grids, Polymers (Basel), 2022, 14:. https://doi.org/10.3390/polym14122376.

[30]   V. Fiore, T. Scalici, G. Di Bella, A. Valenza A review on basalt fibre and its composites, Composites Part B: Engineering, 2015, 74: 74–94. https://doi.org/10.1016/j.compositesb.2014.12.034.

[31]   J.S.S. Neto, R.A.A. Lima, D.K.K. Cavalcanti et al Effect of chemical treatment on the thermal properties of hybrid natural fiber-reinforced composites, Journal of Applied Polymer Science, 2019, 136: 1–13. https://doi.org/10.1002/app.47154.

[32]   N.M. Nurazzi, M.R.M. Asyraf, S. Fatimah Athiyah et al A review on mechanical performance of hybrid natural fiber polymer composites for structural applications, Polymers (Basel), 2021, 13: 1–47. https://doi.org/10.3390/polym13132170.

[33]   Y.G. Thyavihalli Girijappa, S. Mavinkere Rangappa, J. Parameswaranpillai, S. Siengchin Natural fibers as sustainable and renewable resource for development of eco-friendly composites: A Comprehensive review, Frontiers Mater, 2019, 6: 1–14. https://doi.org/10.3389/fmats.2019.00226.

[34]   A. Lotfi, H. Li, D.V. Dao, G. Prusty Natural fiber–reinforced composites: A review on material, manufacturing, and machinability, Journal of Thermoplastic Composite Materials, 2021, 34: 238–284. https://doi.org/10.1177/0892705719844546.

[35]   F. Ahmad, H.S. Choi, M.K. Park A review: Natural fiber composites selection in view of mechanical, light weight, and economic properties, Macromolecular Materials and Engineering, 2015, 300: 10–24. https://doi.org/10.1002/mame.201400089.

[36]   A. Shahzad Hemp fiber and its composites – A review, Journal of Composite Materials, 2012, 46: 973–986. https://doi.org/10.1177/0021998311413623.

[37]   S. Chandgude, S. Salunkhe In state of art: Mechanical behavior of natural fiber-based hybrid polymeric composites for application of automobile components, Polymer Composites, 2021, 42: 2678–2703. https://doi.org/10.1002/pc.26045.

[38]   S.B.R. Devireddy, S. Biswas Physical and Mechanical Behavior of Unidirectional Banana/Jute Fiber Reinforced Epoxy Based Hybrid Composites, Polym Compos, 2017, https://doi.org/10.1002/pc.

[39]   D. Shanmugam, M. Thiruchitrambalam Static and dynamic mechanical properties of alkali treated unidirectional continuous Palmyra Palm Leaf Stalk Fiber/jute fiber reinforced hybrid polyester composites, Materials and Design, 2013, 50: 533–542. https://doi.org/10.1016/j.matdes.2013.03.048.

[40]   N.L. Feng, S.D. Malingam, C.W. Ping, N. Razali Mechanical properties and water absorption of kenaf/pineapple leaf fiber-reinforced polypropylene hybrid composites, Polymer Composites, 2020, 41: 1255–1264.

[41]   M.S. Islam, N.A.B. Hasbullah, M. Hasan et al Physical, mechanical and biodegradable properties of kenaf/coir hybrid fiber reinforced polymer nanocomposites, Materials Today Communications, 2015, 4: 69–76. https://doi.org/10.1016/j.mtcomm.2015.05.001.

[42] R.A. Braga, P.A.A. Magalhaes Analysis of the mechanical and thermal properties of jute and glass fiber as reinforcement epoxy hybrid composites, Materials Science and Engineering: C, 2015, 56: 269–273. https://doi.org/10.1016/j.msec.2015.06.031.

[43] V.P. Arthanarieswaran, A. Kumaravel, M. Kathirselvam Evaluation of mechanical properties of banana and sisal fiber reinforced epoxy composites: Influence of glass fiber hybridization, Materials and Design, 2014, 64: 194–202. https://doi.org/10.1016/j.matdes.2014.07.058.

[44] M. Ramesh, K. Palanikumar, K.H. Reddy Influence of fiber orientation and fiber content on properties of sisal-jute-glass fiber-reinforced polyester composites, Journal of Applied Polymer Science, 2016, 133: 1–9. https://doi.org/10.1002/app.42968.

[45] M. Ramesh, K. Palanikumar, K.H. Reddy Mechanical property evaluation of sisal-jute-glass fiber reinforced polyester composites, Composites Part B: Engineering, 2013, 48: 1–9. https://doi.org/10.1016/j.compositesb.2012.12.004.

[46] E. Selver, N. Ucar, T. Gulmez Effect of stacking sequence on tensile, flexural and thermomechanical properties of hybrid flax/glass and jute/glass thermoset composites, Journal of Industrial Textiles, 2018, 48: 494–520. https://doi.org/10.1177/1528083717736102.

[47] A.A. Pérez-Fonseca, J.R. Robledo-Ortíz, D.E. Ramirez-Arreola et al Effect of hybridization on the physical and mechanical properties of high density polyethylene-(pine/agave) composites, Materials and Design, 2014, 64: 35–43. https://doi.org/10.1016/j.matdes.2014.07.025.

[48] B. Asaithambi, G. Ganesan, S. Ananda Kumar Bio-composites: Development and mechanical characterization of banana/sisal fibre reinforced poly lactic acid (PLA) hybrid composites, Fibers & Polymers, 2014, 15: 847–854. https://doi.org/10.1007/s12221-014-0847-y.

[49] S. Wasti, B. Schwartz, P. Yeole et al Bamboo fiber overmolding textile grade carbon fiber tape and bamboo fiber polypropylene composites, SAMPE Journal, 2023, 59, 22–29.

[50] S. Kore, R. Spencer, H. Ghossein et al Performance of hybridized bamboo-carbon fiber reinforced polypropylene composites processed using wet laid technique, Composites Part C: Open Access, 2021, 6: 100185. https://doi.org/10.1016/j.jcomc.2021.100185.

[51] L. Liu, C. Jia, J. He et al Interfacial characterization, control and modification of carbon fiber reinforced polymer composites, Composites Science & Technology, 2015, 121: 56–72. https://doi.org/10.1016/j.compscitech.2015.08.002.

[52] C.H. Lee, A. Khalina, S.H. Lee Importance of interfacial adhesion condition on characterization of plant-fiber-reinforced polymer composites: A review, Polymers (Basel), 2021, 13: 1–22. https://doi.org/10.3390/polym13030438.

[53] K. Joseph, K. Oksman, G. Gejo et al Surface Treatments in Fiber-reinforced Composites, Elsevier Science & Technology, United Kingdom, 2021, pp. 47–81.

[54] J.M. Stickel, M. Nagarajan Glass fiber-reinforced composites: From formulation to application, International Journal of Applied Glass Science, 2012, 3: 122–136. https://doi.org/10.1111/j.2041-1294.2012.00090.x.

[55] M.R. Zakaria, H. Md Akil, M.H. Abdul Kudus et al Hybrid carbon fiber-carbon nanotubes reinforced polymer composites: A review, Composites Part B: Engineering, 2019, 176: 107313. https://doi.org/10.1016/j.compositesb.2019.107313.

[56] H. Zheng, W. Zhang, B. Li et al Recent advances of interphases in carbon fiber-reinforced polymer composites: A review, Composites Part B: Engineering, 2022, 233: 109639. https://doi.org/10.1016/j.compositesb.2022.109639.

[57] X. Gao, J.W. Gillespie, R.E. Jensen et al Effect of fiber surface texture on the mechanical properties of glass fiber reinforced epoxy composite, Composites: Part A Applied Science and Manufacturing, 2015, 74: 10–17. https://doi.org/10.1016/j.compositesa.2015.03.023.

[58] J. Parameswaranpillai, J.A. Gopi, S. Radoor et al Turning waste plant fibers into advanced plant fiber reinforced polymer composites: A comprehensive review, Composites Part C: Open Access, 2023, 10: 100333. https://doi.org/10.1016/j.jcomc.2022.100333.

[59] K.R. Sumesh, V. Kavimani, G. Rajeshkumar et al Effect of banana, pineapple and coir fly ash filled with hybrid fiber epoxy based composites for mechanical and morphological study, Journal of Material Cycles and Waste Management, 2021, 23: 1277–1288. https://doi.org/10.1007/s10163-021-01196-6.

[60] N.L. Feng, S.D. Malingam, C.W. Ping, N. Razali Mechanical properties and water absorption of kenaf/pineapple leaf fiber-reinforced polypropylene hybrid composites, Polymer Composites, 2020, 41: 1255–1264. https://doi.org/10.1002/pc.25451.

[61] S.K. Saw, G. Sarkhel, A. Choudhury Preparation and characterization of chemically modified jute–coir hybrid fiber reinforced epoxy novolac composites, Journal of Applied Polymer Science, 2012, 125: 3038–3049. https://doi.org/10.1002/app.36610.

[62] T.P. Sathishkumar, P. Navaneethakrishnan, S. Shankar, J. Kumar Mechanical properties of randomly oriented snake grass fiber with banana and coir fiber-reinforced hybrid composites, Journal of Composite Materials, 2013, 47: 2181–2191. https://doi.org/10.1177/0021998312454903.

[63] M.A. AlMaadeed, R. Kahraman, P. Noorunnisa Khanam, N. Madi Date palm wood flour/glass fibre reinforced hybrid composites of recycled polypropylene: Mechanical and thermal properties, Materials and Design, 2012, 42: 289–294. https://doi.org/10.1016/j.matdes.2012.05.055.

[64] M. Valente, F. Sarasini, F. Marra et al Hybrid recycled glass fiber/wood flour thermoplastic composites: Manufacturing and mechanical characterization, Composites: Part A Applied Science and Manufacturing, 2011, 42: 649–657. https://doi.org/10.1016/j.compositesa.2011.02.004.

[65] M. Ghasemzadeh-Barvarz, C. Duchesne, D. Rodrigue Mechanical, water absorption, and aging properties of polypropylene/flax/glass fiber hybrid composites, Journal of Composite Materials, 2015, 49: 3781–3798. https://doi.org/10.1177/0021998314568576.

[66] K. Jarukumjorn, N. Suppakarn Effect of glass fiber hybridization on properties of sisal fiber-polypropylene composites, Composites Part B: Engineering, 2009, 40: 623–627. https://doi.org/10.1016/j.compositesb.2009.04.007.

[67] C. Kahl, M. Feldmann, P. Sälzer, H.P. Heim Advanced short fiber composites with hybrid reinforcement and selective fiber-matrix-adhesion based on polypropylene – Characterization of mechanical properties and fiber orientation using high-resolution X-ray tomography, Composites: Part A Applied Science and Manufacturing, 2018, 111: 54–61. https://doi.org/10.1016/j.compositesa.2018.05.014.

[68] B. Vijaya Ramnath, V.M. Manickavasagam, C. Elanchezhian et al Determination of mechanical properties of intra-layer abaca-jute-glass fiber reinforced composite, Materials and Design, 2014, 60: 643–652. https://doi.org/10.1016/j.matdes.2014.03.061.

[69] F. Pinto, L. Boccarusso, D. De Fazio et al Carbon/hemp bio-hybrid composites: Effects of the stacking sequence on flexural, damping and impact properties, Composite Structures, 2020, 242: 112148. https://doi.org/10.1016/j.compstruct.2020.112148.

[70] S.M. Izwan, S.M. Sapuan, M.Y.M. Zuhri, A.R. Mohamed Thermal stability and dynamic mechanical analysis of benzoylation treated sugar palm/kenaf fiber reinforced polypropylene hybrid composites, Polymers (Basel), 2021, 13, https://doi.org/10.3390/polym13172961.

[71] A. Atiqah, M. Jawaid, S.M. Sapuan et al Thermal properties of sugar palm/glass fiber reinforced thermoplastic polyurethane hybrid composites, Composite Structures, 2018, 202: 954–958. https://doi.org/10.1016/j.compstruct.2018.05.009.

[72] P. Senthil Kumar, K. Gurusami, R. Rajaprasanna et al Characterizations on hybrid of caustic treated natural and glass fiber composites, Materials Today: Proceedings, 2020, 33: 4424–4427. https://doi.org/10.1016/j.matpr.2020.07.669.

[73] M. Ramesh, R. Logesh, M. Manikandan et al Mechanical and water intake properties of banana-carbon hybrid fiber reinforced polymer composites, Materials Research, 2017, 20: 365–376. https://doi.org/10.1590/1980-5373-MR-2016-0760.

[74]   C. Naga Kumar, M.N. Prabhakar, S.J. il Effect of interface in hybrid reinforcement of flax/glass on mechanical properties of vinyl ester composites, Polymer Testing, 2019, 73: 404–411. https://doi.org/10.1016/j.polymertesting.2018.12.005.

[75]   E. Selver Acoustic properties of hybrid glass/flax and glass/jute composites consisting of different stacking sequences, Tekstil Ve Muhendis, 2019, 26: 42–51. https://doi.org/10.7216/1300759920192611305.

[76]   M.S. Kim, J. Yan, K.M. Kang et al Soundproofing properties of polypropylene/clay/carbon nanotube nanocomposites, Journal of Applied Polymer Science, 2013, 130: 504–509. https://doi.org/10.1002/app.39194.

[77]   M.S. Kim, J. Yan, K.H. Joo et al Synergistic effects of carbon nanotubes and exfoliated graphite nanoplatelets for electromagnetic interference shielding and soundproofing, Journal of Applied Polymer Science, 2013, 130: 3947–3951. https://doi.org/10.1002/app.39605.

[78]   A. Haris, U. Kureemun, L.Q.N. Tran, H.P. Lee Noise reduction capability of hybrid flax fabric-reinforced polypropylene-based composites, Advanced Composite Materials, 2019, 28: 335–346. https://doi.org/10.1080/09243046.2018.1533347.

Sangita Tripathy, S. R. Dhakate, Bhanu Pratap Singh

3 Carbon nanomaterials for advanced hybrid composites: properties and applications

Abstract: The never-ending desire of modern humans to meet their basic and advanced needs, along with more comfort in day-to-day life, leads to the application of advanced hybrid materials made up of polymer composites for diverse applications. The reinforcement of one or more filler materials into a thermoplastic or thermosetting polymer matrix forms polymer composites, where the individuals retain their physical properties, but, combined, results in synergistic properties. The carbon nanocomposites are thus multiphase materials consisting of polymer(s) as the matrix and carbon nanomaterials as the reinforcements. Glass, carbon, and/or aramid fibers are sometimes added to provide continuous reinforcement to the nanocomposite system. Recent advances in technologies for the extraction of raw materials and processing of their composites/hybrids help in their cost reduction, thus facilitating their applications in electronics, biomedicine, automobiles, defense, civil aviation, smart fabrics, etc. The excellent capability of carbon nanomaterials, leading to their applications in diverse fields in polymer composites, is discussed here in detail.

3.1 Introduction

An ever-growing population, a continuously changing climate, along with complex international and regional situations in the twenty-first century, motivates the world scientific community to look out for advanced hybrid materials to meet the day-to-day and advanced needs of modern humans, as well as strengthen the defense systems across the world. Polymer composites possess the potential to match the requirements of lightweight, affordability, easy processability, and good thermal and electrical insulating properties required in electronics, aerospace, defense, biomedicals, smart fabrics, etc. [1]. A suitable reinforcement of different weight percentages of metal, ceramic, and/or carbon-based fillers into the polymer matrix enriches the composites, with desirable properties of both polymers and reinforced fillers [2]. There are various methods for intermixing the components of polymer composites, viz. manual mixing, melt-mixing, laser sintering, solvent casting, in situ polymerization, compression molding etc., where each method influences the surface and interface morphology, as well as the physical properties of the

Sangita Tripathy, S. R. Dhakate, Bhanu Pratap Singh, Academy of Scientific and Innovative Research (AcSIR), Ghaziabad 201002, Uttar Pradesh, India; CSIR-National Physical Laboratory, New Delhi 110012, Delhi, India

https://doi.org/10.1515/9783111019543-003

composites differently [3]. The nanofillers, due to their nano-scale size and high surface-to-volume ratio, can introduce desired properties into the polymer composites [4]. Carbon-based fillers, such as carbon nanotubes (CNTs), graphene, fullerene, carbon nanofibers, graphene nanoribbons, etc., occupy a major position among the fillers community, owing to their ability to influence the thermal, mechanical, and electrical properties of the composites with low reinforcements [5, 6]. The combination of polymer matrix (polyethylene, polyamide, polyurethane, polyester, epoxy, etc.) with carbon, glass, aramid fiber, etc. (microscale fillers) and CNTs, graphene, etc. (nanofillers) results in hierarchical or continuous reinforced structures, which have shown directional enhancement in physical properties [7]. The hybridization of different layers of polymers and organic and inorganic materials with metallic or carbon filler reinforcements has shown advanced structural, mechanical, electrical, and thermal properties, and are thus preferred in electronics, biomedicals, energy storage devices, aerospace industry, etc. (Figure 3.1).

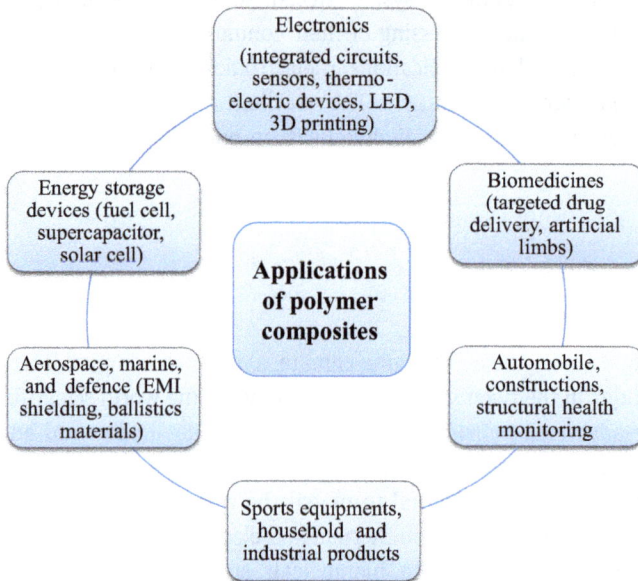

Figure 3.1: Some major applications of polymer composites.

3.2 Types of reinforcements

Reinforcements are basically of two types: (i) discontinuous reinforcement, and (ii) continuous reinforcement. They are determined by the filler's ability to induce directionality in the physical properties of the composites.

Fillers can also be classified as isotropic and anisotropic, based on their ability to align randomly or in a pattern, inside the matrix. Particulate fillers such as carbon black, ceramics, metal oxides, silica, etc. are isotropic as they have a low aspect ratio (length-to-diameter ratio) and their physical properties vary uniformly in all directions [8]. Carbon nanofillers like CNTs, graphene, graphene nanoribbon, carbon nanofibers, etc. can induce directionality with their reinforcements into composites due to their high aspect ratios [8]. Anisotropic fillers can influence the physical properties with minor reinforcements and balance the overall toughness and stiffness of the composites [9].

3.2.1 Discontinuous reinforcement

The random alignment of fillers inside the polymer matrix favors isotropic physical properties, i.e., the mechanical, thermal, or electrical properties do not vary significantly along different directions in a polymer composite. This assures uniformity in physical properties, which is desired in major polymeric applications. The polymer matrix in discontinuous reinforcement can be either thermoplastic or thermosetting, which accordingly gives rise to thermoplastic or thermosetting polymer composites.

The uniform dispersion of anisotropic fillers (fillers with a high aspect ratio) inside a thermoplastic or thermoset matrix results in discontinuously reinforced polymer composites. They have vast applications in the fields of biomedicals, engineering, energy storage, EMI shielding, bio-sensing, etc. [10]. The long carbon, glass, or aramid fibers-reinforced thermoplastics have applications in the automotive industry, military, sports equipment, sailboats, computer hardware, etc., due to their high pressure, temperature, and corrosion endurance [11].

3.2.1.1 Discontinuously reinforced CNTs/polymer composites

CNTs are hollow cylindrical structures of hexagonally arranged carbon atoms, with lengths up to several hundred micrometers and diameters of 1–100 nm [6]. Thus, they can be considered one-dimensional quantum wires with excellent mechanical strength (elastic modulus = 1 TPa and tensile strength of 11–63 GPa) and flexibility [12, 13], thermal conductivity (3,000 W/mK), and electrical conductivity (200,000 S/cm for MWCNTs) [14]. They are basically of two types: single-walled carbon nanotubes (SWCNTs) and multi-walled carbon nanotubes (MWCNTs). The sp^2 hybridization in C–C bonds, high aspect ratio, and high surface area of CNTs allow them to act as excellent reinforcements in thermoplastic polymers for high-strength and flexible applications in automobiles, aerospace, ballistics materials, optoelectronics, strain sensors, etc. [13, 15].

There have been various studies carried out on CNT-reinforced polymer composites by several research groups. Singh et al. utilized the excellent electrical and mechanical properties of MWCNTs to enhance those of MWCNT/polyimide composites, prepared by the solvent-casting method. They reported an enhancement by eighteen orders in the electrical conductivity of 5 wt% amine-functionalized MWCNT/polyimide over that of neat polyimide, whereas the tensile strength and Young's modulus increased by 50% and 35%, respectively (Figure 3.2a) [16]. Gupta et al. observed about thirteen orders of increment in electrical conductivity for 10 wt% MWCNT-reinforced polyurethane composites, along with improved EMI shielding properties [17]. Taher et al. reported an optimum tensile strength of 74.89 MPa and a maximum tensile modulus of 2.91 GPa for 1.8 and 5 wt% MWCNT-reinforced polyamide-6 composites, respectively, prepared by the 3D laser printing method for different structural applications [18]. The tensile strength for 2 wt% MWCNT/polypropylene and 1 wt% CNT/polyamide composites increased by 17.4% and 33.33%, respectively, as reported by Mei et al. [19].

The MWCNTs, due to their high thermal conductivity and mechanical strength, have been preferred as reinforcements in polymers (high-density polyethylene (HDPE), polyethylene terephthalate (PET), polymethyl methacrylate (PMMA), etc.) for the fabrication of aerospace components [20]. Li et al. reported enhanced proton transmission capacity, reduced secondary neutrons, and high thermal stability in CNTs/PMMA composites designed for radiation shielding applications in outer space [21]. The MWCNTs-reinforced metal matrix composites have shown highly improved toughness, fatigue properties, and tensile properties, hence preferred for the fabrication of lightweight body parts in aircraft. Shin and Bae reinforced Al2024 with 0.3–4 vol% MWCNTs, and they reported a 175% increase in fatigue strength and a 101.4% increase in the ultimate strength of 4 vol% MWCNTs/Al2024 composites [22]. The long polymeric structures, consisting of repeating units of C–C chains in CNTs, provide them with utmost flexibility, thus they can act as shock absorbers in ballistic materials [23]. The mechanics of CNTs in ballistic applications have been suitably exploited, with their continuous (secondary) reinforcements in thermoplastic/thermoset matrix/glass or aramid fiber composites, which will be discussed in a later part.

The high electrical conductivity in CNTs imparts high sensing ability to mechanical damages in CNT-reinforced cement or concrete-based constructions, whereas the self-healing nature of CNTs promises smart infrastructures for the future [24]. The utilization of smart materials to understand the periodic response of a mechanical system, as a response to an external force, is studied under structural health monitoring (SHM). Dinesh et al. prepared different wt% of MWCNT-reinforced cement-based strain sensors, which showed good mechanical strength, electrical conductivity, and piezo-resistive properties, required in SHM [25].

Polymers are viscous materials. CNTs, due to their excellent mechanical properties, facilitate a temperature- and strain rate-dependent viscous-to-elastic transition in polymer composites. The reinforcement of CNTs also increases the glass transition temperature (Tg) of polymers, making them useful for designing automotive, electronics, and household products [26]. A temperature-dependent viscoelastic transition was observed at 5 vol% of CNTs-reinforced 50 wt/wt% PA12/PA6 composites prepared by

Clemente et al. by twin screw extrusion [27]. Moud et al. studied the rheological and electrical properties of polyamide/polypropylene blend, reinforced with 1–5 wt% of MWCNTs, and they noticed an increasing trend in electrical conductivity, as well as storage and loss modulus values for increasing reinforcements [28]. Jyoti et al. reported a significant rise in storage modulus, loss modulus, and flexural modulus of MWCNTs, GO, and GO-MWCNTs-reinforced ABS composites, with increasing reinforcements up to 10 wt% [29]. Sethy et al. reported polymeric glass transition temperature-dependent electrical conductivity in MWCNTs/PA-12/PP composites, which were suitable for co-continuous structured thermoelectric devices [30].

CNTs-reinforced composites have applications in real-time sensing of signals coming from biological cells, owing to their high aspect ratio, fast electron transfer kinetics, and chemical functionalization with most of the available chemical species. CNTs-reinforced polymer-based FETs, chemiresistors, and electrodes have shown very good electrochemical sensing properties [31, 32]. The large surface area, hollow cylindrical geometry, chemical activeness to functional groups, and high electrical conductivity of CNTs make excellent gas sensors out of CNTs/polymer composites. CNTs play a significant role in toxic gas detection in the atmosphere, such as CO, NO_2, SO_2, etc. [33]. Kumar et al. have reported 37% sensitivity to NO_2 gas by polyethylenimine-functionalized SWCNTs, fabricated on a SiO_2 substrate, as compared to only 20.12% sensitivity by pristine SWCNTs/SiO_2 [34]. Singh et al. reported continuously improving electrical conductivity with increasing reinforcements for 0.1–0.5 wt% short- and long-length MWCNTs in epoxy composites (Figures 3.2b and 2c) [35]. Chauhan et al. studied the dynamic mechanical properties of 1, 2, 3, and 5 wt% MWCNTs-reinforced poly(ether) ketone (PEK) composites, prepared by melt-mixing, and they reported 19% and 89% improvements in the storage modulus of the composites in the glassy and rubbery regions, respectively, for 5 wt% MWCNTs/PEK, as shown in Figure 3.2d [36]. A maximum 29% improvement in the tensile strength was observed for 3 wt% MWCNTs-reinforced ABS composites (Figure 3.2e), whereas the electromagnetic shielding effectiveness showed a continuously increasing trend with increasing reinforcements (Figure 3.2f) [37].

3.2.1.2 Discontinuously reinforced graphene/polymer composites

Graphene is defined as a single-layered sheet of hexagonally bonded (honeycomb structure) carbon atoms. It is termed a 2D nanomaterial due to the confinement of electron flow along the x-y plane. The lightweight, high surface-to-volume ratio, good thermal conductivity (~5,000 W/mK), high electron mobility (200,000 cm^2/V-s), and high mechanical strength (130 GPa) properties make graphene and graphene oxide (GO) the primary choice among carbon nanofillers for multiple applications [6], which is why it has largely taken over the global market in the last decade [38]. Graphene, graphene oxide (GO), or graphene nanoplate (GNP)-reinforced thermoplastic composites find diverse applications, such as mechanical reinforcements for aero-

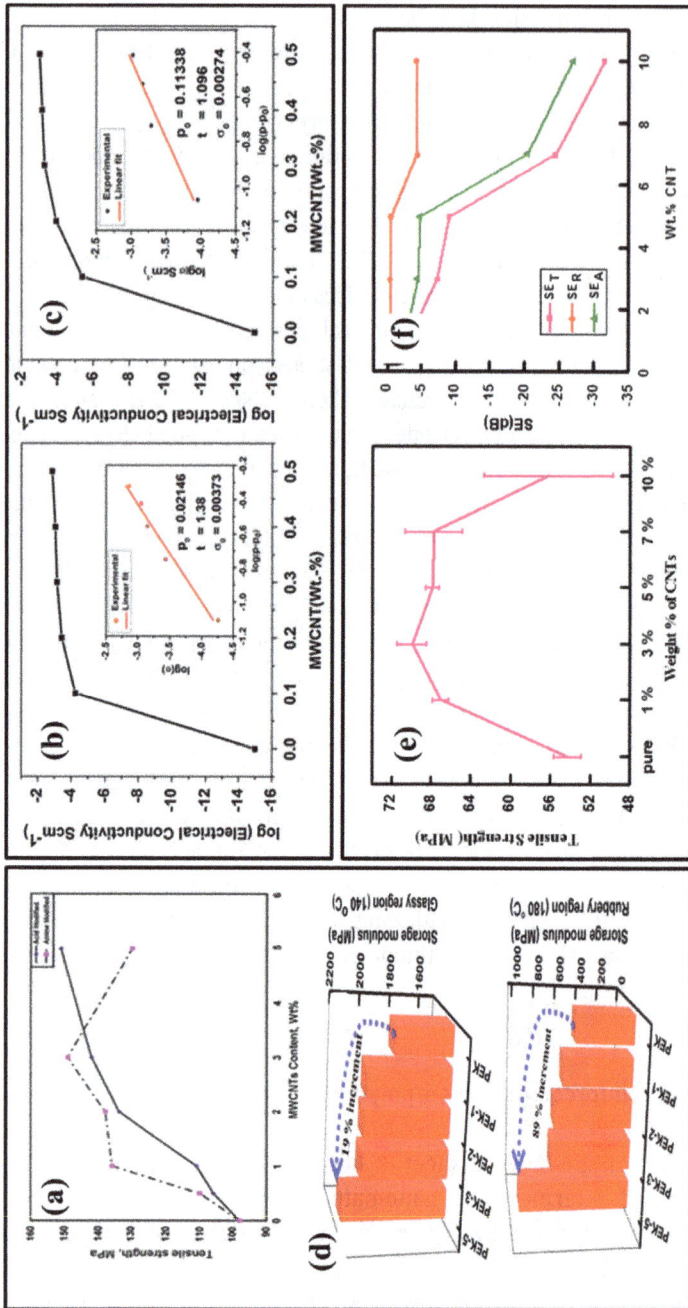

Figure 3.2: (a) Tensile strength for varying acid-functionalized and amine-functionalized MWCNTs in a polyimide matrix [16] (reproduced with permission from Springer, copyright © 2008), variation of electrical conductivity for (b) long-length MWCNTs-reinforced epoxy composites, and (c) short-length MWCNTs-reinforced epoxy composites [35](reproduced with permission from Springer, copyright © 2014), (d) Storage modulus in the glassy and rubbery regions in DMA measurement for varying MWCNT amounts in poly(ether)ketone [36] (reproduced with permission from Wiley, copyright © 2018), (e) variation of tensile strength, and (f) variation of EMI shielding effectiveness, for varying MWCNT amounts in ABS composites [37] (reproduced with permission from Elsevier, copyright © 2015).

space and automobile parts, carriers and sensors in biomedicine, gas sensors, electronics and optoelectronics, smart and light fabrics, conducting polymer composites, self-healing materials, shape memory composites, water purification, etc. [8, 39]

The aromatic C–C chains in graphene provide it with superior mechanical strength, which can be further improved with functionalization, acid treatment, etc., in order to enhance the mechanical properties of polymer composites [6]. Tripathy et al. reinforced 0.1–0.5 parts per hundred ratios (phr) GO into polyamide-6 by melt-mixing. The composite specimens prepared by the injection molding process showcased significantly improved tensile as well as thermal properties at 0.1 phr loading. Figures 3.3a and 3.3b show the FESEM images for as-synthesized GO sheets and the tensile-tested fractured surface of 0.5 phr GO/PA6 composite [40]. Gupta et al. reported enhanced nano-mechanical properties in 0–5 wt% rGO-reinforced polyurethane (PU) composites. Their hardness and elastic modulus (measured by the nanoindentation method) increased by 139% and 129%, respectively, for 5 wt% reinforcement, which showcased their potential for scratch-resistant automobile coatings. Figure 3.3c shows the FESEM images of rGO. The hardness and elastic modulus graphs for the rGO/PU composites are shown in Figure 3.3d [41]. Zhou et al. reported significantly improved tensile strength (155% improvement for 0.5 wt% reinforcement) and glass transition temperature (5.5 °C improvement for 1 wt% reinforcement) for fluorine group functionalized graphene/polyimide composites, for friction-proof applications in automobiles, aerospace, packaging, etc. [42]. Kiziltas et al. melt-mixed up to 8 wt% GNPs into bio-based polyamide-6,10, and they reported a maximum of twofold enhancement in tensile modulus with 8 wt% GNPs reinforcement [43].

The ultralightweight of graphene sheets, along with high electron mobility, allows their applications in human motion monitoring, artificial muscle design, chemical sensing, gas sensing, etc. [44]. Choi et al. reported very good gas sensitivity of graphene sheets fabricated on copper substrate by CVD, owing to the high surface-to-volume ratio of graphene [45]. Chun et al. observed accurate strain sensing, up to 0.1%, shown by CVD-grown graphene sheets, later transferred to a polydimethylsiloxane (PDMS) substrate [46].

The oxygen-containing functional groups like –OH, -COOH, epoxy, etc., attached to atomically thin layers of graphene, provide them with very good antimicrobial and permeation properties, which have been further improved by their attachment with metallic molecules, biomolecules, and polymers [47]. Alammar et al. prepared polybenzimidazole (PBI)-GO composite membranes for effective oil-water separation, which showed a permeance of 91.3 $L/m^2/h/bar$ and an efficiency of 99.9%, along with good antibacterial properties [48]. The nanofiltration membranes made of suitably intercalated GO-phytic acid (PhA) composites (1:10), fabricated by Zhu et al., showed a pure water flux of 6.31 $L/m^2/h$ and a rejection rate of 99.88%, along with high pH balance, which was quite improved compared to other composite filtration membranes [49]. Sharma et al. prepared rGO, molybdenum silicate (MoS_2), and rGO/MoS_2-reinforced TPU composites by solution mixing, and they observed 39% and 140% improvements in specific shielding effectiveness of rGO/MoS_2/TPU composites over that of rGO and MoS_2 composites, respectively (Figure 3.3e).

Also, the rGO/MoS$_2$/TPU composites showed an improved tensile strength (increased by 23%) compared to neat TPU specimens (Figure 3.3f) [50]. Verma et al. prepared barium ferrite-decorated rGO composite by ball milling to explore their EMI shielding properties (Figure 3.3g) and the composite showed a total shielding effectiveness of 32 dB in the 12.4–18 GHz frequency range of the Ku-band (Figure 3.3h) [51].

Graphene-polymer composites are widely used as biosensors, owing to the high surface area, high electron mobility, good thermal, electrical, and optical properties, and high flexibility of graphene [52].

Moreover, diverse applications of graphene and graphene oxide have been reported in different fields, and researchers are still trying to utilize the ultrahigh thermal, electrical, and mechanical properties of this unique nanomaterial in polymer composites.

3.2.1.3 Discontinuously reinforced graphene nanoribbon/polymer composites

Graphene nanoribbons are thin and long strips of hexagonally arranged carbon atoms [53]. Several methods have been adopted for their synthesis, such as lithography, plasma etching, chemical vapor deposition, oxidative unzipping, etc. [54]. The graphene oxide nanoribbons (GONR) are chemically synthesized by oxidative unzipping of multi-walled carbon nanotubes (MWCNTs) in the presence of oxidizing agents and strong acids [53]. A schematic diagram of CNTs' unzipping is shown in Figure 3.4a. The GONR are chemically or thermally reduced into graphene nanoribbons (GNR) in order to utilize their superior electrical properties in nanocomposites [54, 55]. A good mechanical reinforcement inside the polymer matrix is assured by GONR reinforcements due to the attached functional groups at the surfaces and edges. The superior mechanical, thermal, electrical, and viscoelastic properties of GNR-reinforced polymer composites with lower reinforcements inside the matrix have enabled their applications in fuel cells, supercapacitors, EMI shielding, gas sensors, etc. [54, 56]. Jun et al. compared the reinforcement of MWCNTs and GNR into a TPU matrix separately, prepared by solution mixing and compression molding. They observed a total shielding effectiveness of 24.9 dB in 8.2 vol % GNR/TPU, as compared to 9.3 dB, with an equal amount of MWCNTs reinforcement (Figure 3.4b). They also observed higher strength values in GNR/TPU composites over a similar amount of MWCNTs-reinforced TPU composites, attributed to the large surface area and presence of active reaction sites at the edges in GNR (Figure 3.4c) [55].

The large surface area of GNR, with attached functional groups at the surfaces and edges (hydroxyl (-OH), carbonyl (-CO-), and carboxyl (-COOH), etc.), assures better interfacial bonding with the polymer matrix, thus enhancing the mechanical properties at comparatively lower reinforcements. Rafiee et al. observed 22% and 30% improvements in ultimate tensile strength and Young's modulus values, respectively, for 0.3 wt% GNR/epoxy composites, as compared to MWCNTs/epoxy composites, which showed only 2–4% improvement in strength and modulus [57]. Shang et al. reinforced 1 and 2 wt%

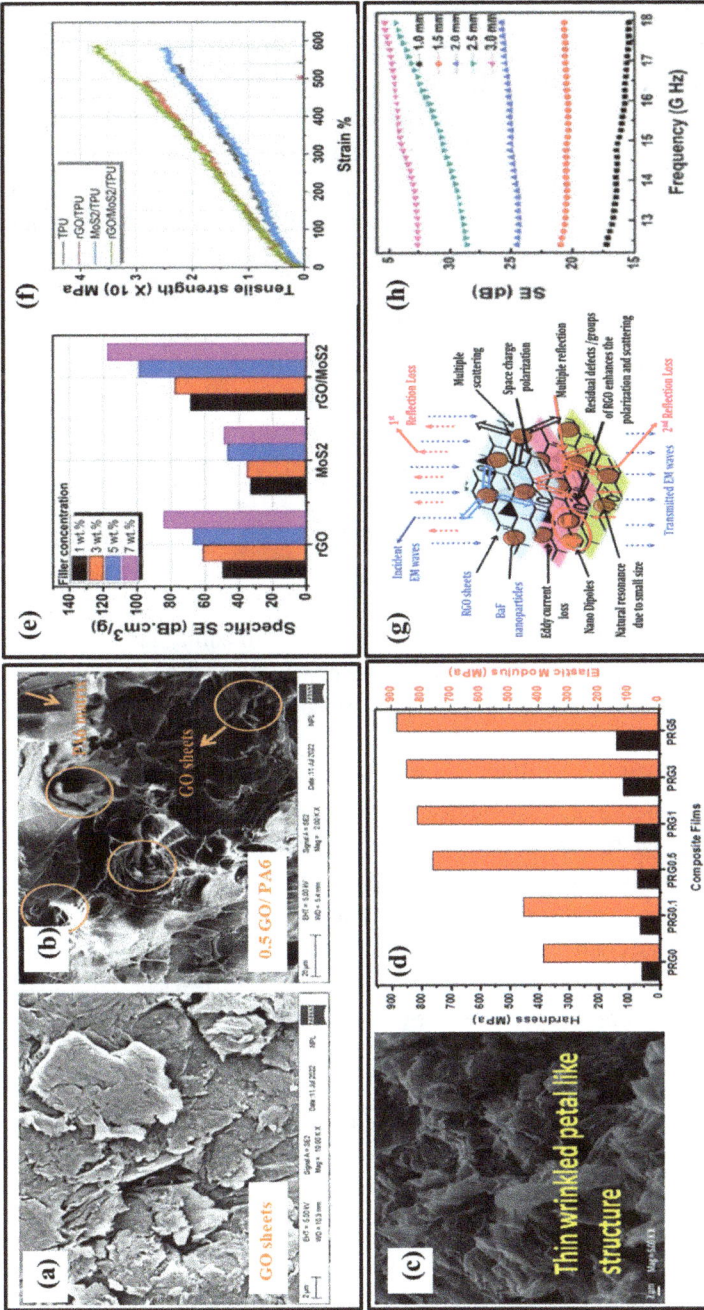

Figure 3.3: (a) FESEM image of GO sheets, and (b) fractured surface FESEM image of 0.5 phr GO/polyamide-6 [40] (reproduced with permission from Wiley, copyright © 2023) (c) SEM image of rGO, (d) hardness and elastic modulus of rGO/PU films with varying rGO amounts [41] (reproduced with permission from RSC, copyright © 2015), (e) EMI shielding effectiveness, and (f) tensile strength values for varying nanofiller amounts in rGO/MoS₂/polyurethane composite foam [50] (reproduced with permission from Elsevier, copyright © 2023), (g) EMI shielding mechanism in rGO/barium ferrite composites, (h) total shielding effectiveness in rGO/barium ferrite films for varying film thickness [51] (reproduced with permission from RSC, copyright © 2015).

GNR into polyvinyl alcohol (PVA), and they noticed 85.7% and 65.2% increments in tensile strength and Young's modulus, respectively, for 2 wt% GNR reinforcement. These composites were designed for high-strength polymeric applications [58]. Yang et al. showed a comparative study on the reinforcing effect of GNR and MWCNTs in PVA by the solution mixing method. The tensile strength and Young's modulus for 3.6 vol% GNR/PVA increased by 17% and 32.5%, respectively, over 3.6 vol% MWCNTs/PVA. The planar structure and large surface area with many attached functional groups, unlike the limited surface area of MWCNTs, maximized the polymer–filler–filler interactions. The electrical conductivity also increased by six orders of magnitude for GNR reinforcements, which was nearly an order higher than that for MWCNTs reinforcement (Figure 3.4d) [59]. Liu et al. reinforced 0.1–3 wt% GNR into a polyamide matrix by the solution mixing method. They observed a 38% improvement in tensile strength and a maximum of 50% improvement in elongation with just 0.1 wt% reinforcement. The mechanical properties further showed a decreasing trend at higher loadings due to the agglomeration of nanoribbons (Figure 3.4e), whereas the electrical resistivity continuously decreased with increasing reinforcements. Indicated better electron transport in the composites with increasing GNR content (Figure 3.4f) [60]. Habibpour et al. prepared 0.5–2 wt% MWCNTs, GNR, and graphene nanoplates (GnP), separately reinforced into polyurethane (PU) composites by the solvent casting method. The GNR/PU composites showed significantly higher Young's modulus and tensile strength values compared to those of MWCNTs/PU and GnP/PU composites. A higher interaction among the large surface area of GNR and the PU matrix was responsible for this [61].

The excellent electrical conductivity of GNR has been utilized in GNR-reinforced epoxy composites for generating the Joule heating effect, which has applications in the aircraft industry to remove moisture from the aircraft exterior. This removal is crucial, as moisture would otherwise increase the weight and consume more fuel. Raji et al. reported 1 S/cm conductivity shown by 5 wt% GNR-reinforced epoxy specimens, which were able to generate 0.5 W/cm^2 power density [62]. Polymer–GNR composites have shown good response to low-concentration toxic gases like CO, NH$_3$, N$_2$O, etc., due to the large surface area of GNR and proper polymer-GNR interactions, as observed by Trajcheva et al. They used methyl methacrylate, butyl acrylate, and 2-hydroxyethyl methacrylate monomers and reinforced 0.2–3 wt% GNR into them by an in situ polymerization process [63]. The reduced GONR-PU 3D nanoporous foam structures with PDMS coating have shown good compressive strength and electrical conductivity, owing to the large surface area of GONR, and hence used for effective oil/water separations [64].

3.2.1.4 Discontinuously reinforced CNTs/graphene/carbon fiber hybrid/polymer composites

The individual reinforcement of CNTs and graphene sheets, though showing enhanced physical properties in composites, often presents challenges in their processing due to

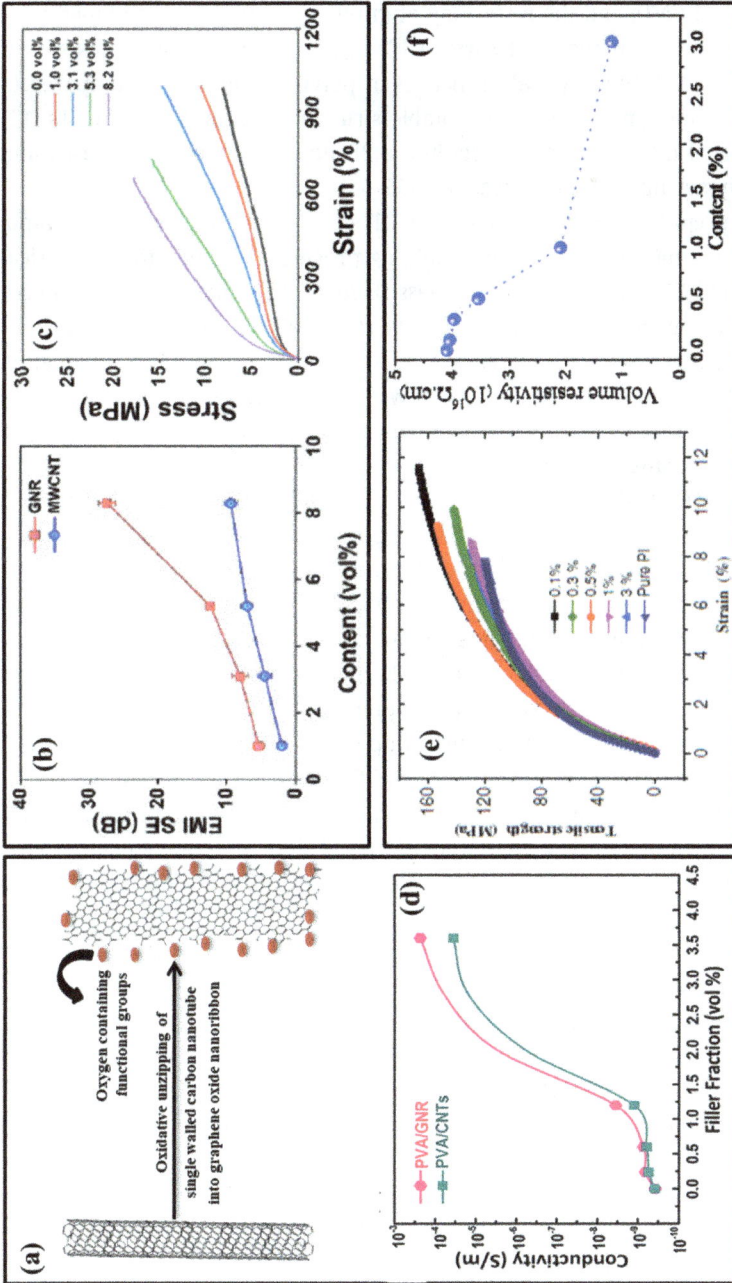

Figure 3.4: (a) Schematic diagram of chemical unzipping of typical SWCNT into GONR with attached functional groups, (b) total EMI shielding effectiveness for varying GNR and CNTs vol% separately in TPU composites, (c) stress-strain curves for varying GNR vol% in TPU composites [55] (reproduced with permission from Elsevier, copyright © 2021), (d) electrical conductivity for varying GNR and CNTs reinforcements separately in PVA composites [59] (reproduced with permission from Springer Nature, copyright © 2017), (e) stress-strain curves, and (f) volume resistivity for varying amounts of reinforcement of GNR within a polyimide matrix [60] (reproduced with permission from Elsevier, copyright © 2016).

the agglomeration of CNTs and the restacking of graphene sheets. The combined reinforcement of CNTs and graphene as hybrids (GCNTs), on the other hand, has shown overall improvements in the physical properties of polymer composites compared to their individual reinforcements [65]. The suitable structural combination of CNTs (1D) and two-dimensional (2D) graphene sheets into a 3D nanostructure can interact and bond with the polymer matrix more effectively [66–66].

Verma et al. prepared 0.5–10 wt% rGO-MWCNTs-reinforced polyurethane composites by solution mixing and compression molding processes to study their electrical conductivity (σ) and EMI shielding effectiveness (Figure 3.5a). They observed a maximum shielding effectiveness of – 47 dB in 10 wt% rGO-MWCNTs/PU (Figure 3.5b), and the σ continuously increased with increasing reinforcements (Figure 3.5c). Zhao et al. studied the EMI shielding properties of 0.05–8 wt.% CNTs, 0.2–15 wt% graphene, and 5 wt% CNTs/10 wt% graphene-reinforced poly(vinylidene fluoride) (PVDF) composites. They reported an EMI shielding effectiveness of 27.58 dB in graphene/CNTs/PVDF composites, which was 23% higher than CNTs/PVDF and 9.3% higher than graphene/PVDF. The composite films can be used for ultrathin, flexible, and conductive electronic components [69]. The GCNT hybrids-reinforced thermoplastic polyurethane (TPU) films have shown superior mechanical strength and toughness over neat TPU films, as reported by Li et al. They observed a 1.9 times improvement in tensile strength and a 2.9 times improvement in toughness values for 1 wt% reinforcement. The composite films promised excellent potential for flexible sensor applications [68]. Dey et al. studied the thermal, mechanical, and electrical properties of up to 10 wt% GO, CNTs, and GO/CNTs-reinforced polybenzimidazole composites, and the GO/MWCNTs (1:1)/PBI composites exhibited the highest tensile strength of 2.94 GPa among the composites, which was 90% and 119%, respectively, higher than those for GO and MWCNTs-reinforced composites. The synergistic effect of GO/MWCNTs hybrids also enhanced the dc conductivity of composites by six orders as compared to neat PBI [70].

Babal et al. reported a 40.75% improvement in tensile strength and a 203.3% improvement in the storage modulus of 2 wt% COOH-functionalized MWCNTs/18 wt% carbon fibers (chopped)-reinforced polycarbonate composites, prepared by melt-mixing and injection molding, which were designed for high-strength applications. The synergistic effect of both fibers and MWCNTs induced superior mechanical properties in the composites [71]. Jyoti et al. studied the detailed static and dynamic mechanical properties of 1, 3, 5, 7, and 10 wt% MWCNTs-, rGO-, and rGO-MWCNTs-reinforced ABS composites. They reported a maximum tensile strength of 62.1 MPa (26.1% improvement) in 7 wt% rGO-MWCNTs/ABS composites, whereas the Young's modulus continuously enhanced with reinforcements for all composites. Their detailed tensile behavior is shown in Figure 3.5d–3.5i [72].

We discussed several types of continuous reinforcements and their overall impact on the physical properties of polymer composites. However, some limitations have been observed in terms of effective load transfer from the polymer matrix to the carbon nanofillers. The probability of agglomeration and restacking in these nanofillers

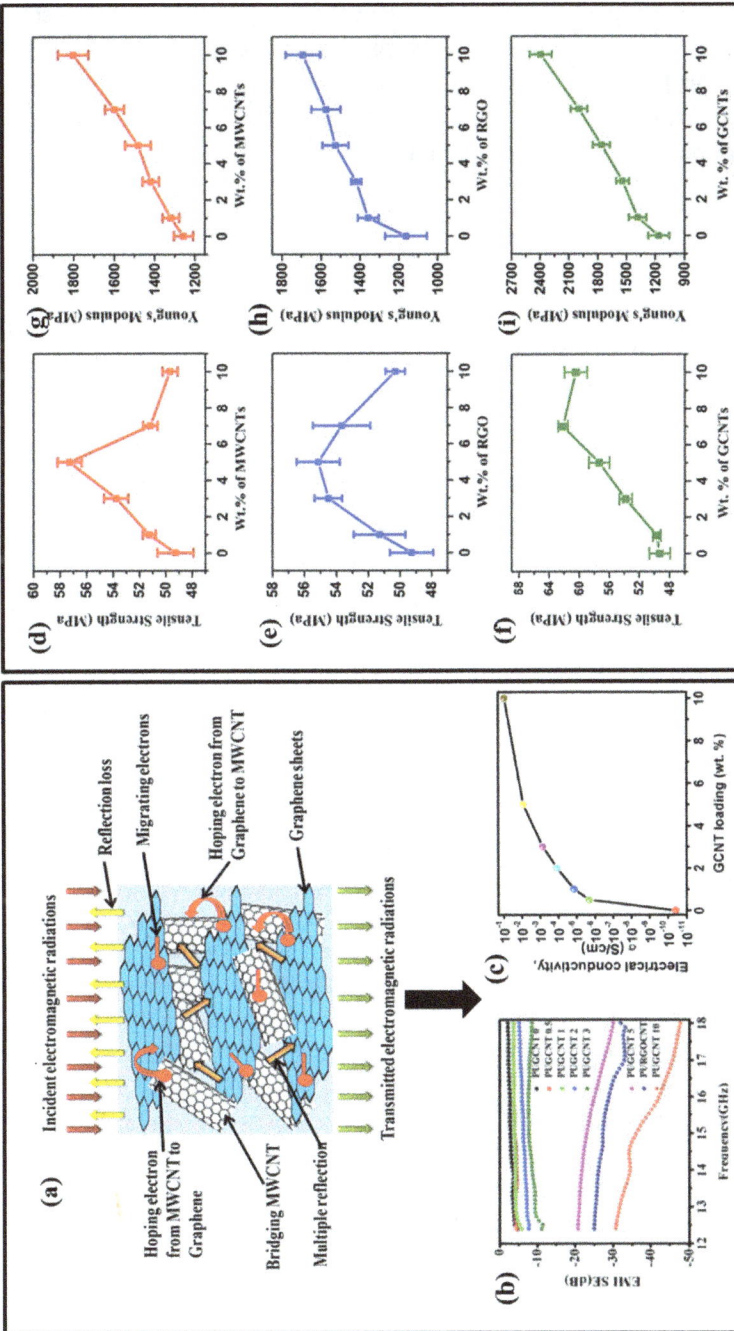

Figure 3.5: (a) Schematic image of GCNTs-PU interactions for EMI shielding, (b) EMI shielding effectiveness with varying frequency in the Ku band for GCNTs-PU composites, (c) Electrical conductivity for varying GCNTs reinforcement in PU composites [67]. (reproduced with permission from Elsevier, copyright © 2017) (d-f) Tensile strength for varying wt% of reinforcements of MWCNTs, reduced GO, and GCNTs in ABS matrix, (g-i) Young's modulus for varying wt% of reinforcements of MWCNTs, reduced GO, and GCNTs in ABS matrix [72]. (reproduced with permission from Springer, copyright © 2018).

does not allow complete exploration of the excellent thermal, mechanical, and electrical properties of these nanomaterials, which overly compromises the material properties of the composites [73]. Thus, there was a need to address the limitations in these nanomaterials by combining them with micro-sized fillers, which can prevent agglomerations and encourage better interfacial bonding.

3.2.2 Continuously reinforced carbon/polymer composites

The carbon nanofillers like CNTs, graphene, graphene nanoribbons, etc., when reinforced together with microscale carbon fibers (CF) into a thermoplastic or thermoset polymer matrix, are called hierarchical or continuous reinforcements. The carbon nanofillers-CF network fosters both in-plane and out-of-plane mechanical properties in the composites manyfold [7]. The high strength and stability of continuous reinforcement systems in nature, such as plant cell walls and animal skeletons, inspired researchers to work on hybridized continuous reinforcements in laboratories [74]. The 0.1 wt% functionalized-GO/CNTs-reinforced PA-6,6/CF composites showed 6.1 °C and 9.3 °C enhancements in melting temperature and crystallization temperature, respectively, as reported by Cho et al. The tensile strength and storage modulus values of the composite specimen improved by 136% and 300%, respectively. Moreover, the superior mechanical and thermal properties of the composite can be attributed to the successful formation of a hybrid nanofillers network inside the PA-6,6 matrix [75].

Sharma et al. synthesized 0.05–5 wt% MWCNTs-reinforced Kevlar (an aramid, i.e., aromatic polyamide fiber)/epoxy hybrid composites by solvent casting and compression molding method (Figure 3.6a). They reported a 91% improvement in tensile strength (Figure 3.6b), a 233% increment in storage modulus (Figure 3.6c), a 33% increment in flexural modulus, and a 50% increment in Young's modulus, respectively, for 0.3 wt% MWCNTs/Kevlar/epoxy hybrids, as compared to neat kevlar fiber. The hybrids altogether promised excellent mechanical properties, required for ballistics applications [76]. The hybrid reinforcement of carbon fibers and MWCNTs into epoxy matrix by Singh et al. provided a good reinforcement (SEM image shown in Figure 3.6d), which resulted in a continuously increasing electrical conductivity (σ) with increasing reinforcements, as shown in Figure 3.6e [77]. Sharma et al. later prepared hybrid GO-CNTs-reinforced polycarbonate/Kevlar composites, which showed very good static and dynamic mechanical properties, required in strong, protective, and flexible body armors. The storage modulus, tensile strength, and Young's modulus of 0.2 wt% GO-CNTs-reinforced composite tapes showed 188%, 32%, and 44% improvements, respectively.(Figure 3.6f–3.6 h) [78].

Lian et al. reinforced 0.1–2 wt% of Kevlar-GNR (KGNR), GNR, MWCNTs, and Kevlar-CNTs (KCNTs) into polyvinyl chloride (PVC) and poly-methyl methacrylate (PMMA) matrices by the solution-mixing method. They observed a 127% improvement in tensile strength for 0.3 wt% KGNR/PMMA and a 106% improvement in yield strength for 1 wt%

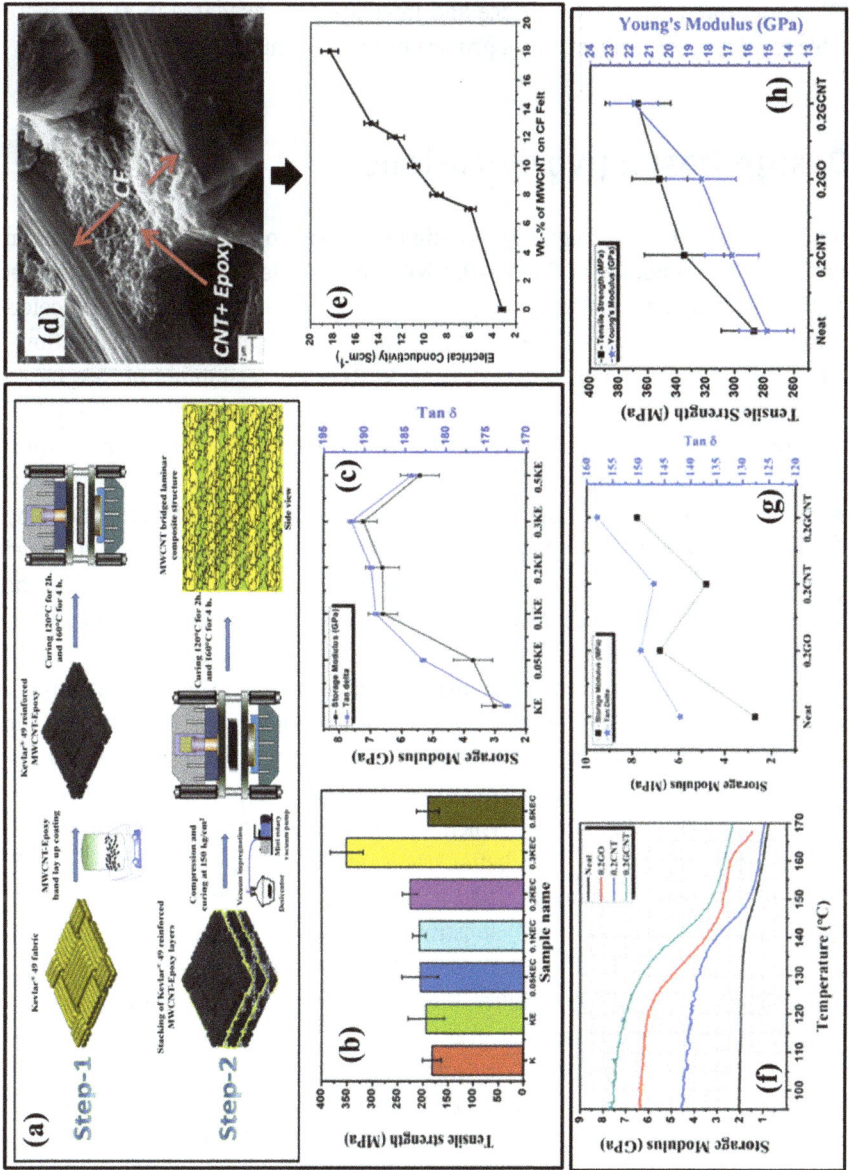

Figure 3.6: (a) Schematic diagram of MWCNTs-reinforced aramid-epoxy composite tape preparation by compression molding technique, (b) variation of tensile strength, and (c) storage modulus and loss factor, for varying MWCNT amounts inside aramid-epoxy composites [76] (reproduced with permission from Elsevier, copyright © 2018), (d) fracture surface image of MWCNT-CF-epoxy composites, and (e) electrical conductivity for varying MWCNT amounts inside CF-epoxy composites [77] (reproduced with permission from Springer, copyright © 2014), (f-h) variation of storage modulus, tanδ, tensile strength, and Young's modulus of GO, CNT, and GCNT-reinforced kevlar/polycarbonate composites [78] (reproduced with permission from Elsevier, copyright © 2020).

KGNR/PVC composites. The active surface and interface area available in KGNRs were responsible for the superior mechanical properties of the composites [79].

3.3 Design-based hybridizations

The interface of multilayered metallic/ceramic and/or carbon-based systems is often doped with requisite amounts of CNTs, GO, GNR, etc., in order to achieve desired mechanical, thermal, and electrical properties in solar cells, photovoltaic cells, thermoelectric devices, biomedicines, etc. They can be termed as design-based hybridizations. Mohammadnezhad et al. explored the role of MWCNTs in the stability of TiO_2-based dye-sensitized solar cells. The MWCNTs-reinforced TiO_2 composites offered a directional path to the photo-injected electrons due to their superior electrical conductivity; thus, the lifetime of electrons was enhanced and the carrier recombination rate was reduced, which enhanced the shelf-life of the solar cells. The MWCNTs partially absorbed and acted as a blocking agent for the UV light, thereby preventing degradation, as observed from UV spectroscopy measurements. The dye desorption was also decreased by the MWCNTs reinforcement, as confirmed from Raman spectroscopic measurements. They reported an improved current density and thermal stability in the TiO_2 solar cells with MWCNTs doping, which promised dye-sensitized solar cells with long-term stability [80]. Chakraborty et al. studied the effect of GO with varying content deposited on top of fluorine-doped tin oxide (SnO_2) coated glass substrates for dye-sensitive solar cells. They observed a visible enhancement in current density and power conversion efficiency (η) with increasing GO content, which was attributed to reduced recombination of photo-induced charge carriers with better transport, facilitated by GO networks [81]. As per Zulkifli et al., graphene layers acted as a suitable interface for Gallium nitride (GaN) Nano rods grown on silicon (111) substrates for photo-detection applications. The limitations of low electrical and thermal conductivity, rigidity, opacity, etc., of silicon were fulfilled by the high carrier mobility, thermal conductivity, and wide spectral absorption wavelength of graphene; thus, graphene acted as a perfect carrier transport channel in GaN-based detectors. The GaN/Graphene/Si devices showed a photo responsivity of 17.4 A/W and specific detectivity of 1.23×10^{13} Jones [82]. Graphene sheets also acted as perfect sinks in metal–insulator–superconductor junctions due to their high thermal and electrical conductivity; thus, a fast responsivity of 50 nA/pW and noise equivalent power of 10^{-18} W/(Hz)$^{-0.5}$ were reported by Vischi et al. with their fast bolometer model [83]. Lamichhane et al. successfully fabricated MWCNTs-MgB$_2$ yarns by coating MgB$_2$-isopropanol solution on vertically aligned CVD-grown MWCNTs. The yarns showed very good tensile properties (strength = 200 MPa, Young's modulus = 1.27 GPa) and current density ($J_C = 3.4 \times 10^7$ A/cm^2), with a superconducting transition temperature of 38 K, thus having promising applications in high-power electronics [84]. Zheng et al. fabricated multilayers of MWCNTs-reinforced TPU and polycaprolactone composites by multiple extrusions, designed for tunable thermo and electro-responsive shape-memory effects. The eight-layered composites showcased very good conductivity along with fast electro response, whereas, a maximum yield strength of 35.6 MPa was shown by the composite with 32 layers. Thus, MWCNTs promised excellent reinforcements for shape-memory polymer devices [85]. As per a study carried out by Bharti et al., the conversion of p-type charge carriers in MWCNTs bucky papers into n-type on polyethylenimine (PEI) treatment allowed 227 µV/7.6 µA output current for a 40 °C temperature difference. Moreover, the MWCNTs provided a cheaper and environmentally friendly option for the fabrication of thermoelectric devices [86].

3.4 Conclusion and future scopes

The superior physical properties of carbon nanomaterials have allowed their application in polymer composites in diverse fields such as modern electronics, optoelectronics, defense and aerospace equipment, automobile components, energy storage devices, biomedical devices, and water filtration. The hybridization of CNTs, graphene sheets, etc., with microscale carbon or glass fibers further enhances their in-plane and out-of-plane mechanical properties, which have shown promising applications for ballistics components, aerospace components, etc. The suitable integration of carbon nanomaterials with polymer or metallic frameworks has shown promising applications in solar cells, supercapacitors, thermoelectric devices, and superconducting devices. They altogether point to a new era of science dominated by carbon allotropes in the coming decades.

References

[1]   B. Hosnedlova *et al.* Carbon nanomaterials for targeted cancer therapy drugs: A critical review, The Chemical Record, 2019, 19: 502–522.
[2]   M. Xanthos Polymers and polymer composites, Functional Fillers for Plastics, 2005, 1: 1–16.
[3]   H.V. Madhad, D.V. Vasava Review on recent progress in synthesis of graphene–polyamide nanocomposites, Journal of Thermoplastic Composite Materials, 2022, 35: 570–598.
[4]   A. Thabet, Y. Mubarak, M.A. Bakry review of nano-fillers effects on industrial polymers and their characteristics, Journal of Engineering Science, 2011, 39: 377–403.
[5]   I.O. Oladele, T.F. Omotosho, A.A. Adediran Polymer-based composites: An indispensable material for present and future applications, International Journal of Polymer Science, 2020, 2020: 1–12.
[6]   R.B. Mathur, B.P. Singh, S. Pande Carbon Nanomaterials: Synthesis, Structure, Properties and Applications, Taylor & Francis, New York, 1st Ed. 2016.
[7]   A.N.M.M.R. Abdullah Sayam, M. Sakibur Rahman et al. A review on carbon-fiber reinforced polymer composites: Mechanical performance, manufacturing process, structural applications and allied challenges, Carbon Letters, 2022, 32: 1173–1205.
[8]   A. Bandyopadhyay, P. Dasgupta, S. Basak Engineering of Thermoplastic Elastomer with Graphene and Other Anisotropic Nanofillers, Springer Singapore, 1st Ed. 2020.

[9]   M. Staropoli *et al.* Hybrid silica-based fillers in nanocomposites: Influence of isotropic/isotropic and isotropic/anisotropic fillers on mechanical properties of styrene-butadiene (SBR)-based rubber, Polymers, 2021, 13: 2413.

[10] M. Nofar, R. Salehiyan, S.S. Ray Influence of nanoparticles and their selective localization on the structure and properties of polylactide-based blend nanocomposites, Composites Part B: Engineering, 2021, 215: 108845.

[11] H. Ning et al. A review of long fibre thermoplastic (LFT) composites, International Materials Reviews, 2020, 65: 164–188.

[12] J.N. Coleman, U. Khan, W.J. Blau, Y.K. Gun'ko Small but strong: A review of the mechanical properties of carbon nanotube–polymer composites, Carbon, 2006, 44: 1624–1652.

[13] N. Gupta, S.M. Gupta, S. Sharma Carbon nanotubes: Synthesis, properties and engineering applications, Carbon Letters, 2019, 29: 419–447.

[14] B.P. Singh, K.M. Subhedar, Emerging Applications of Carbon Nanotubes and Graphene, 2023.

[15] N. Nurazzi *et al.* Mechanical performance and applications of CNTs reinforced polymer composites – A review, Nanomaterials, 2021, 11: 2186.

[16] B. Singh, D. Singh, R. Mathur, T. Dhami Influence of surface modified MWCNTs on the mechanical, electrical and thermal properties of polyimide nanocomposites, Nanoscale Research Letters, 2008, 3: 444–453.

[17] T. Gupta et al. Designing of multiwalled carbon nanotubes reinforced polyurethane composites as electromagnetic interference shielding materials, Journal of Polymer Research, 2013, 20: 1–7.

[18] F. Taher, H. Hardani, S. Bakhshi, M.R. Samadi, M. Ayaz Enhancing the tensile properties of PA6/CNT nanocomposite in selective laser sintering process, Polymer Composites, 2023, 44 (2): 1290–1304.

[19] X. Mei, D. Ye, F. Zhang, C.A. Di Implantable application of polymer-based biosensors, Journal of Polymer Science, 2022, 60: 328–347.

[20] A. Bhat et al. Review on nanocomposites based on aerospace applications, Nanotechnology Reviews, 2021, 10: 237–253.

[21] Z. Li *et al.* PMMA/MWCNT nanocomposite for proton radiation shielding applications, Nanotechnology, 2016, 27: 234001.

[22] S. Shin, D. Bae Fatigue behavior of Al2024 alloy-matrix nanocomposites reinforced with multi-walled carbon nanotubes, Composites Part B: Engineering, 2018, 134: 61–68.

[23] S.-Y. Wu, P. Sikdar, G.S. Bhat Recent progress in developing ballistic and anti-impact materials: Nanotechnology and main approaches, Defence Technology, 2023, 21: 33–61.

[24] M. Siahkouhi, G. Razaqpur, N. Hoult, M.H. Baghban, G. Jing Utilization of carbon nanotubes (CNTs) in concrete for structural health monitoring (SHM) purposes: A review, Construction and Building Materials, 2021, 309: 125137.

[25] A. Dinesh, S. Sudharsan, S. Haribala Self-sensing cement-based sensor with carbon nanotube: Fabrication and properties–A review, Materials Today: Proceedings, 2021, 46: 5801–5807.

[26] W.K. Goertzen, M. Kessler Dynamic mechanical analysis of carbon/epoxy composites for structural pipeline repair, Composites Part B: Engineering, 2007, 38: 1–9.

[27] L. Arboleda-Clemente, X. García-Fonte, M.-J. Abad, A. Ares-Pernas Role of rheology in tuning thermal conductivity of polyamide 12/polyamide 6 composites with a segregated multiwalled carbon nanotube network, Journal of Composite Materials, 2018, 52: 2549–2557.

[28] A. Abbasi Moud, A. Javadi, H. Nazockdast, A. Fathi, V. Altstaedt Effect of dispersion and selective localization of carbon nanotubes on rheology and electrical conductivity of polyamide 6 (PA 6), Polypropylene (PP), and PA 6/PP nanocomposites, Journal of Polymer Science Part B: Polymer Physics, 2015, 53: 368–378.

[29] J. Jyoti, S. Dhakate, B.P. Singh Phase transition and anomalous rheological properties of graphene oxide-carbon nanotube acrylonitrile butadiene styrene hybrid composites, Composites Part B: Engineering, 2018, 154: 337–350.

[30] S. Sethy, V. Barwal, B.K. Satapathy Tunable thermo-sensitive electrical conductivity of melt-mixed PA-12/PP-MWCNT nanocomposites, Composites Science and Technology, 2022, 217: 109099.
[31] K. Balasubramanian, M. Burghard Biosensors based on carbon nanotubes, Analytical and Bioanalytical Chemistry, 2006, 385: 452–468.
[32] H. Gergeroglu, S. Yildirim, M.F. Ebeoglugil Nano-carbons in biosensor applications: An overview of carbon nanotubes (CNTs) and fullerenes (C 60), SN Applied Sciences, 2020, 2: 1–22.
[33] P. Dariyal, S. Sharma, G.S. Chauhan, B.P. Singh, S.R. Dhakate Recent trends in gas sensing via carbon nanomaterials: Outlook and challenges, Nanoscale Advances, 2021, 3: 6514–6544.
[34] S. Kumar, V. Pavelyev, P. Mishra, N. Tripathi Thin film chemiresistive gas sensor on single-walled carbon nanotubes-functionalized with polyethylenimine (PEI) for NO _ 2 NO 2 gas sensing, Bulletin of Materials Science, 2020, 43: 1–7.
[35] B. Singh et al. Effect of length of carbon nanotubes on electromagnetic interference shielding and mechanical properties of their reinforced epoxy composites, Journal of Nanoparticle Research, 2014, 16: 1–11.
[36] S.S. Chauhan, B.P. Singh, R.S. Malik, P. Verma, V. Choudhary Detailed dynamic mechanical analysis of thermomechanically stable melt-processed PEK–MWCNT nanocomposites, Polymer Composites, 2018, 39: 2587–2596.
[37] J. Jyoti, S. Basu, B.P. Singh, S. Dhakate Superior mechanical and electrical properties of multiwall carbon nanotube reinforced acrylonitrile butadiene styrene high performance composites, Composites Part B: Engineering, 2015, 83: 58–65.
[38] S.K. Tiwari, R.K. Mishra, S.K. Ha, A. Huczko Evolution of graphene oxide and graphene: From imagination to industrialization, ChemNanoMat, 2018, 4: 598–620.
[39] X. Fu et al. Graphene oxide as a promising nanofiller for polymer composite, Surfaces and Interfaces, 2023, 37: 102747.
[40] S. Tripathy et al. Mechanical and thermal properties of polyamide-6 nanocomposites reinforced by graphene oxide with low loading by double extrusion method, Polymers for Advanced Technologies, 2023, 34 (8): 2788–2798.
[41] T.K. Gupta et al. Superior nano-mechanical properties of reduced graphene oxide reinforced polyurethane composites, Rsc Advances, 2015, 5: 16921–16930.
[42] S. Zhou et al. Tribological behaviors of polyimide composite films enhanced with fluorographene, Colloids and Surfaces A: Physicochemical and Engineering Aspects, 2019, 580: 123707.
[43] A. Kiziltas, W. Liu, S. Tamrakar, D. Mielewski Graphene nanoplatelet reinforcement for thermal and mechanical properties enhancement of bio-based polyamide 6, 10 nanocomposites for automotive applications, Composites Part C: Open Access, 2021, 6: 100177.
[44] A. Kausar Trends in graphene reinforced polyamide nanocomposite for functional application: A review, Polymer-Plastics Technology and Materials, 2019, 58: 917–933.
[45] J.H. Choi et al. Graphene-based gas sensors with high sensitivity and minimal sensor-to-sensor variation, ACS Applied Nano Materials, 2020, 3: 2257–2265.
[46] S. Chun, Y. Choi, W. Park All-graphene strain sensor on soft substrate, Carbon, 2017, 116: 753–759.
[47] X. Zhang et al. Graphene-based functional hybrid membranes for antimicrobial applications: A review, Applied Sciences, 2022, 12: 4834.
[48] A. Alammar, S.-H. Park, C.J. Williams, B. Derby, G. Szekely Oil-in-water separation with graphene-based nanocomposite membranes for produced water treatment, Journal of Membrane Science, 2020, 603: 118007.
[49] L. Zhu et al. Graphene oxide composite membranes for water purification, ACS Applied Nano Materials, 2022, 5: 3643–3653.
[50] S. Sharma, B.P. Singh, S.H. Hur, W.M. Choi, J.S. Chung Facile fabrication of stacked rGO/MoS2 reinforced polyurethane composite foam for effective electromagnetic interference shielding, Composites Part A: Applied Science and Manufacturing, 2023, 166: 107366.

[51]   M. Verma *et al.* Barium ferrite decorated reduced graphene oxide nanocomposite for effective electromagnetic interference shielding, Physical Chemistry Chemical Physics, 2015, 17: 1610–1618.

[52]   W.K. Abdelbasset *et al.* Comparison and evaluation of the performance of graphene-based biosensors, Carbon Letters, 2022, 32: 927–951.

[53]   D.V. Kosynkin *et al.* Longitudinal unzipping of carbon nanotubes to form graphene nanoribbons, Nature, 2009, 458: 872–876.

[54]   A. Kausar Graphene nanoribbon: Fundamental aspects in polymeric nanocomposite, Polymer-Plastics Technology and Materials, 2019, 58: 579–596.

[55]   Y.-S. Jun et al. Enhanced electrical and mechanical properties of graphene nano-ribbon/thermoplastic polyurethane composites, Carbon, 2021, 174: 305–316.

[56]   M.D. Prasad, A. Sharma, P. Tambe In Journal of Physics: Conference Series, IOP Publishing, p. 012004.

[57]   M.A. Rafiee et al. Graphene nanoribbon composites, ACS Nano, 2010, 4: 7415–7420.

[58]   S. Shang, L. Gan, C.W.M. Yuen, S.-X. Jiang, N.M. Luo The synthesis of graphene nanoribbon and its reinforcing effect on poly (vinyl alcohol), Composites Part A: Applied Science and Manufacturing, 2015, 68: 149–154.

[59]   M. Yang *et al.* Simultaneously improving the mechanical and electrical properties of poly (vinyl alcohol) composites by high-quality graphitic nanoribbons, Scientific Reports, 2017, 7: 17137.

[60]   X. Liu et al. Dielectric and mechanical properties of polyimide composite films reinforced with graphene nanoribbon, Surface and Coatings Technology, 2017, 320: 497–502.

[61]   S. Habibpour et al. Structural impact of graphene nanoribbon on mechanical properties and anti-corrosion performance of polyurethane nanocomposites, Chemical Engineering Journal, 2021, 405: 126858.

[62]   A.-R.O. Raji et al. Composites of graphene nanoribbon stacks and epoxy for joule heating and deicing of surfaces, ACS Applied Materials & Interfaces, 2016, 8: 3551–3556.

[63]   A. Trajcheva et al. QCM nanocomposite gas sensors–Expanding the application of waterborne polymer composites based on graphene nanoribbon, Polymer, 2021, 213: 123335.

[64]   C.-F. Cao *et al.* Design of mechanically stable, electrically conductive and highly hydrophobic three-dimensional graphene nanoribbon composites by modulating the interconnected network on polymer foam skeleton, Composites Science and Technology, 2019, 171: 162–170.

[65]   J. Jyoti, B.P.A. Singh review on 3D graphene–carbon nanotube hybrid polymer nanocomposites, Journal of Materials Science, 2021, 56: 17411–17456.

[66]   J. Jyoti *et al.* Synergetic effect of graphene oxide-carbon nanotube on nanomechanical properties of acrylonitrile butadiene styrene nanocomposites, Materials Research Express, 2018, 5: 045608.

[67]   M. Verma, S.S. Chauhan, S. Dhawan, V. Choudhary Graphene nanoplatelets/carbon nanotubes/ polyurethane composites as efficient shield against electromagnetic polluting radiations, Composites Part B: Engineering, 2017, 120: 118–127.

[68]   L. Li et al. Molecular-engineered hybrid carbon nanofillers for thermoplastic polyurethane nanocomposites with high mechanical strength and toughness, Composites Part B: Engineering, 2019, 177: 107381.

[69]   B. Zhao, C. Zhao, R. Li, S.M. Hamidinejad, C.B. Park Flexible, ultrathin, and high-efficiency electromagnetic shielding properties of poly (vinylidene fluoride)/carbon composite films, ACS Applied Materials and Interfaces, 2017, 9: 20873–20884.

[70]   B. Dey *et al.* Enhancing electrical, mechanical, and thermal properties of polybenzimidazole by 3D carbon nanotube@ graphene oxide hybrid, Composites Communications, 2020, 17: 87–96.

[71]   A.S. Babal *et al.* Synergistic effect on static and dynamic mechanical properties of carbon fiber-multiwalled carbon nanotube hybrid polycarbonate composites, RSC Advances, 2016, 6: 67954–67967.

[72] J. Jyoti, A.S. Babal, S. Sharma, S. Dhakate, B.P. Singh Significant improvement in static and dynamic mechanical properties of graphene oxide–carbon nanotube acrylonitrile butadiene styrene hybrid composites, Journal of Materials Science, 2018, 53: 2520–2536.

[73] J. Banerjee, K. Dutta Melt-mixed carbon nanotubes/polymer nanocomposites, Polymer Composites, 2019, 40: 4473–4488.

[74] E.S.G. Hui Qian, M.S.P. Shaffer, A. Bismarck Carbon nanotube-based hierarchical composites: A review, Journal of Material Chemistry, 2010, 20: 4751–4762. doi: 10.1039/c000041h.

[75] B.-G. Cho et al. Enhancement in mechanical properties of polyamide 66-carbon fiber composites containing graphene oxide-carbon nanotube hybrid nanofillers synthesized through in situ interfacial polymerization, Composites Part A: Applied Science and Manufacturing, 2020, 135: 105938.

[76] S. Sharma et al. Excellent mechanical properties of long multiwalled carbon nanotube bridged Kevlar fabric, Carbon, 2018, 137: 104–117.

[77] B. Singh et al. Enhanced microwave shielding and mechanical properties of high loading MWCNT–epoxy composites, Journal of Nanoparticle Research, 2013, 15: 1–12.

[78] S. Sharma, J. Rawal, S.R. Dhakate, B.P. Singh Synergistic bridging effects of graphene oxide and carbon nanotube on mechanical properties of aramid fiber reinforced polycarbonate composite tape, Composites Science and Technology, 2020, 199: 108370.

[79] M. Lian, J. Fan, Z. Shi, H. Li, Y.J. Kevlar® -functionalized graphene nanoribbon for polymer reinforcement, Polymer, 2014, 55: 2578–2587.

[80] M. Mohammadnezhad et al. Role of carbon nanotubes to enhance the long-term stability of dye-sensitized solar cells, ACS Photonics, 2020, 7: 653–664.

[81] M. Chakraborty, R. Banerjee, R.N. Gayen AIP Conference Proceedings., AIP Publishing LLC, p. 020062.

[82] N.A.A. Zulkifli et al. A highly sensitive, large area, and self-powered UV photodetector based on coalesced gallium nitride nanorods/graphene/silicon (111) heterostructure, Applied Physics Letters, 2020, 117: 191103.

[83] F. Vischi et al. Electron cooling with graphene-insulator-superconductor tunnel junctions for applications in fast bolometry, Physical Review Applied, 2020, 13: 054006.

[84] U. Lamichhane et al. Twisted laminar superconducting composite: MgB 2 embedded carbon nanotube yarns, Bulletin of Materials Science, 2021, 44: 1–8.

[85] Y. Zheng, B. Zeng, L. Yang, J. Shen, S. Guo Fabrication of thermoplastic polyurethane/ polycaprolactone multilayered composites with confined distribution of MWCNTs for achieving tunable thermo-and electro-responsive shape-memory performances, Industrial & Engineering Chemistry Research, 2020, 59: 2977–2987.

[86] M. Bharti et al. Free-standing flexible multiwalled carbon nanotubes paper for wearable thermoelectric power generator, Journal of Power Sources, 2020, 449: 227493.

Umesh Marathe, Georges Chahine, Sanjita Wasti, Chase Mccullar,
Pritesh Yeole, Sana Elyas, Soydan Ozcan, Uday Vaidya

4 Bridging conventional manufacturing through hybrid manufacturing processes

4.1 Introduction

Hybrid composite materials are gaining traction in various applications like automotive, sports, construction, etc. These materials have multiple advantages, such as reducing embodied carbon and bestowing overall high functionalities. However, hybrid composites are produced using conventional techniques such as compression molding, extrusion, filament winding, and injection molding. Disadvantages of traditional techniques arise due to material-specific selection of processes, cycle time, tooling cost, etc. Figure 4.1 depicts the advantages and disadvantages of preprocessing as per the required matrix forms (powder, solution, fibers, and melts); it lists requirements, advantages, and disadvantages, particularly for fabric-based unidirectional or bidirectional composites.

Apart from processability, the other parameters, such as crystallinity, dispersion, distribution, and orientation of the reinforcing phase, could be affected by processing techniques. Moreover, the parameters mentioned above eventually affect the final performance properties [2, 3].

The modern approach to producing polymer composites, i.e., hybrid composite processing technologies, combines multiple conventional or advanced manufacturing techniques to render hybrid composite processing technologies. The advantages of hybrid processing technologies include saving cycle time per product, near net shape manufacturing, low skilled labor requirements, efficient adaptability, and automated processes. Besides these, hybrid manufacturing processes bestow better-performing

Notice: This manuscript has been authored by UT-Battelle, LLC, under contract DE-AC05-00OR22725 with the US Department of Energy (DOE). The US government retains and the publisher, by accepting the article for publication, acknowledges that the US government retains a nonexclusive, paid-up, irrevocable, worldwide license to publish or reproduce the published form of this manuscript, or allow others to do so, for US government purposes. DOE will provide public access to these results of federally sponsored research in accordance with the DOE Public Access Plan (https://www.energy.gov/doe-public-access-plan).

Umesh Marathe, Sana Elyas, Soydan Ozcan, Manufacturing Science Division, Oak Ridge National Laboratory, Oak Ridge, TN, United States
Georges Chahine, Sanjita Wasti, Chase Mccullar, Pritesh Yeole, Tickle College of Engineering, University of Tennessee, Knoxville, Knoxville, TN, United States
Uday Vaidya, Manufacturing Science Division, Oak Ridge National Laboratory, Oak Ridge, TN, United States; Tickle College of Engineering, University of Tennessee, Knoxville, Knoxville, TN, United States; Institute for Advanced Composites Manufacturing Innovation (IACMI), Knoxville, TN, United States

https://doi.org/10.1515/9783111019543-004

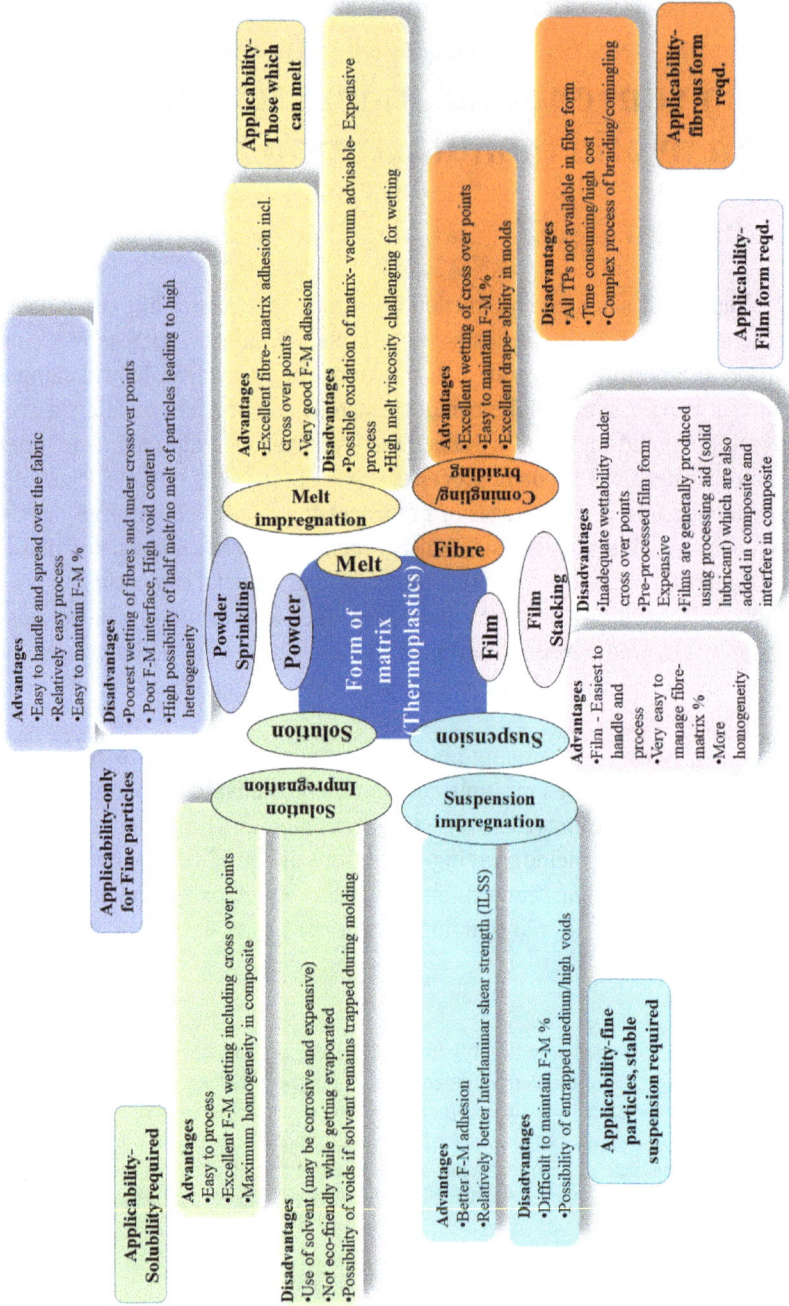

Figure 4.1: Various preprocessing techniques for fabric-reinforced thermoplastic composites (reproduced with permission from Elsevier, copyright © 2021) [1].

composites due to the involvement of more than one process, such as high dispersion and distribution of fibers and fillers, efficient wetting of the reinforcing phase, and fiber orientation.

This chapter briefly overviews conventional processes like compression molding, injection molding, braiding, commingling, automated tape placement, and filament winding and their effect on the performance properties of materials. The chapter also summarizes hybrid composite processing techniques – extrusion compression molding, injection-compression molding, additive manufacturing-compression molding with conventional techniques such as compression molding, injection molding, braiding, commingling, filament winding, automated tape placement, over-molding techniques, and fiber-reinforced metal laminates. Furthermore, it focuses on the utility of various conventional composite processing techniques and their hybrid combinations that enable the fabrication of composite materials with superior properties. It also provides a glimpse of progress in hybrid composite manufacturing.

4.2 Conventional processes

4.2.1 Compression molding

Compression molding is one of the conventional techniques popular in small- and large-scale industries. The tooling and operating costs for compression molding are relatively low compared to other available batch-based polymer processing techniques. Apart from the industries, this technique is popular in composites research. Compression molding is used to manufacture multiple components, such as simple structural parts, household items, tires, automotive parts, and so on [4]. Meliande et al. developed aramid and curaua fiber hybrid composites using cold compression molding. The study assesses the possibility of curaua fibers replacing aramid fibers in ballistic helmet applications. The study concludes that the curaua fibers show potential, yet it is impossible to replace the aramid fibers in ballistic helmet applications [5]. Jiang et al. reported the hybrid composites of carbon fibers and aramid fiber layered assembly. These assemblies were further strengthened using multi-walled carbon nanotubes. The composites were processed by hand layup followed by compression molding (1.5 MPa for 12 h at room temperature). Multi-walled carbon nanotubes were found to prevent crack growth and bestow the bridging effect, suppressing the delamination [6]. Proietti et al. recycled thermoset composites using compression molding. The thermoset composite panels were delaminated using a roll-forming process. The separated plies were compression-molded using polyamide 6 film to demonstrate a circular economy [7]. Marathe et al. used an impregnation-co-film stacking approach to develop polyaryletherketone bidirectional composites with a modified interface with polyetherimide using compression molding. The study comprises the preparation of polyetherimide solution (0, 3, 5, 10, 15, and 20 wt% solution) prepared in dichloromethane. The carbon fabric (twill weave, 385 gsm) was impregnated and dried. The dried fabrics were stacked alternatively with polyaryletherketone films, followed by compression molding. The processes involve impregnation of carbon fiber weave followed by matrix film stacking and eventual compression molding. 10 and 15 wt% of PEI solution-impregnated graphite fabric in polyaryletherketone (PAEK) composite improved interlaminar shear strength by 120% and 130%, respectively [1].

4.2.2 Injection molding

Injection molding is a versatile polymer/composite processing technology known for its high production rate and is widely used for the mass production of various plastic components. There are ample studies reported on the manufacturing of composites using injection molding, the effect on the orientation of composite constituents, the effect on the dispersion of fibers and fillers, etc. Doagoa-Red et al. explored the nano (MWCNTs) and micro (copper microfibers) composites using injection molding. The study was supported using experimental tests and simulations. It was concluded that hybridization of filler led to increased electrical and thermal conductivity by twofold and 37%, respectively [8]. Zhang et al. reported insert molding using injection molding with textured and untextured steel with glass fiber-reinforced polyamide 66 composites. It shows that the steel bears the mechanical load, whereas the failure of the specimen initiates at the steel-composite interface [9]. Carvalho et al. explored polypropylene, short glass fiber, and hollow glass beads ternary composites with the help of injection molding. It also explores the aspects of amino silane treatment to hollow glass beads. This hybrid composition showed improved performance with amino silane treatment [10].

Contrary to the abovementioned compositions, Lee et al. explored the combination of glass/kenaf/bamboo fiber-reinforced polypropylene using injection molding. A composite with 30 wt% kenaf fiber and 10 wt% bamboo showed an improvement of 39% [11].

Bai et al. reported the fabrication of a thermally conductive yet electrically insulative polycarbonate composite using high-shear injection molding. Adding boron nitride particles and multi-walled carbon nanotubes improved thermal conductivity by 832% [12].

4.2.3 Braiding and commingling

Braiding and commingling-based composite manufacturing are unique processing technologies that require matrix materials in fiber form. Braiding involves angular interlacing (more ordered tow) of matrix and reinforcement fibers, whereas, in commingling, monofilaments of matrix and reinforcement are mixed into a single tow. Both techniques are well known for effectively wetting reinforcing fibers in composite fabrication. Marrivada et al. used triaxially braided glass fibers and epoxy resin to fabricate composites. It was argued that adding graphene nanoplatelets to the epoxy improved the interfacial shear strength by 10%, flexural strength by 20%, and tensile strength and modulus by 10% and 30%, respectively. The improvement in strength and suppression of cracks could be attributed to the bridging effect and graphene-induced toughening of epoxy [13]. Gu et al. reported the effect of braiding angle on the interlaminar shear strength for hybrid composites. The study also comprises the fabrication of 3D hybrid composites by introducing a hybrid layer between the Kevlar and carbon fiber regions, whereas in the case of conventional composites, braided layers of carbon and aramid were co-molded (in the absence of the hybrid region). It was found that the hybrid sandwich composite showed better performance compared to the conventional composite. Delamination was the primary failure mechanism for the conventional composite, whereas the 3D hybrid composite showed interlaminar debonding. While analyzing the braiding angle, it was found that the 20° braiding angle shows better performance [14].

Abdkader et al. reported that unidirectional composites were developed using commingled yarn and air texturing. The study involves the design of the air texturing nozzle and hybrid yarn (glass fibers, stainless steel, and polyamide 6 filaments) (Figure 4.2). The outcome shows that the fiber-metal hybrid composite has a tensile strength of 700 MPa and a modulus of 55 GPa [15].

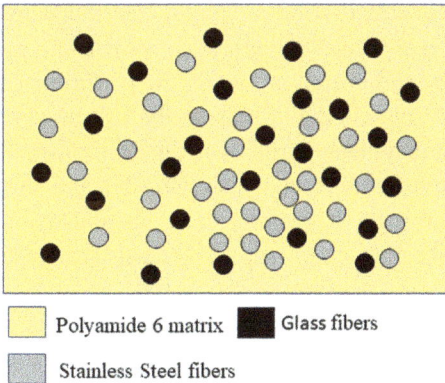

☐ Polyamide 6 matrix ■ Glass fibers
☐ Stainless Steel fibers

Figure 4.2: Schematic representation of a cross section of hybrid composites depicting glass fiber, stainless steel filaments, and a polyamide-6 matrix [15].

Awais et al. compared natural fiber composites (jute, hemp, and flax) fabricated using woven, woven commingled, and knitted commingled fabrics. Woven fabric-based composite was prepared by stacking an alternating sequence of woven matrix fabric. Woven commingled composite was prepared by stacking woven commingled fabric (where the warp was reinforcing yarn and the weft was matrix yarn), and knitted commingled composite was prepared by knitting matrix yarn and reinforcing yarns (matrix vol%: 35). For the short beam test, knitted commingled composite showed excellent performance, especially with flax fibers [16]. Benedetto et al. reported the development of hybrid composites with metallic braided mesh. Five types of metallic meshes were combined with the polyetheretherketone (PEEK)/CF commingled fibers by filament winding followed by compression molding. Combining materials, i.e., braided metal mesh and commingled fibers, improved shear performance [17].

4.2.4 Automated tape placement

Automated tape placement (ATP) is widely used to produce high-end structures for aircraft and other industries where precision process control and quality are paramount. The ATP process consists of three essential components: tape (polymer-impregnated fiber tape), heating system, and compaction mechanism. The tape used for ATP could be thermoset-based or thermoplastic-based matrix materials. The variation in the heating system is usually between a nitrogen-based heat torch integrated with the head and laser-assisted heating. Consolidation parts comprise rollers with different variations in cylindrical geometry. The process involves laying a pre-impregnated unidirectional tape on the mandrel and compacting it against it under the required temperature. The process offers reduced material wastage, allows complex part manufacturing, reduces placement time, is less labor-intensive, and improves the safety of workers, among other benefits. However, it comes with a different set of challenges, such as low inter-ply adhesion, and poor compaction compared to conventional processes like compression molding and autoclave molding. The ATP-manufactured parts are reprocessed using compression or autoclave molding to mitigate these challenges. Qureshi et al. studied the process parameters to assess their effect on the consolidation efficiency and overall impact on the performance properties of produced composites. The study reports the lacunas in the ATP system, which could be mitigated using efficient heat flux systems. Furthermore, it compares nozzle distance, nitrogen rate (L/min), compaction force (lb), tool temperature (°C), and tape layup speed (mm/s) using interlaminar shear strength. It was concluded that adding another process, like compression or autoclave molding, could help further strengthen the composites [18].

Stokes-Griffin and Compston studied various aspects of the AFP process with a laser for carbon/PEEK prepreg tapes [19]. Laminates fabricated using a lower line speed (100 mm/s) exhibited better fiber-matrix bonding than those manufactured using a higher line speed (400 mm/s). The authors attributed this to the higher crystallinity in the laminates fabricated at 100 mm/s. A deformable roller (silicone roller) resulted in better consolidation than a rigid roller (brass roller). This could be attributed to the large laser shadow produced by the rigid roller. For the modeling of the consolidation process, the authors showed that the temperature threshold for adhesion of PEEK to occur is Tg instead of Tm if the PEEK is completely melted during the heating phase. Figure 4.3 depicts the fundamental ATP process.

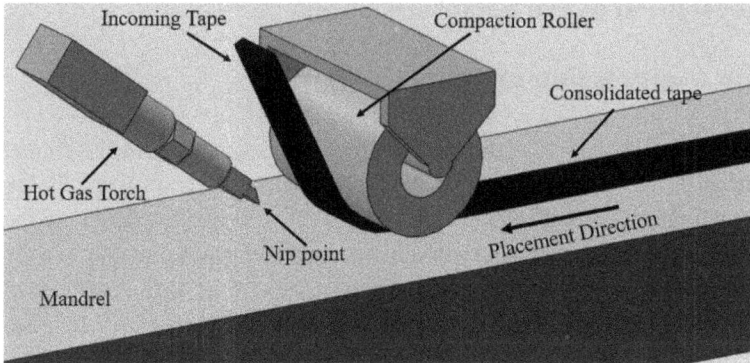

Figure 4.3: ATP schematics for automated tape placement [19].

4.2.5 Filament winding

Filament winding is a popular composite manufacturing technique for hollow cylindrical parts. The process comprises winding continuous fiber tows on a rotating mandrel at a certain angle. Parts produced using filament winding are used in the aerospace, energy, and automotive industries. Since it can produce hollow tanks and pressurized vessels, it attracts attention for fuel storage applications. Filament-wound parts are usually cured in an oven or autoclave system. There is ample research available on filament winding-based composite materials [20].

Pandita et al. reported a novel clean filament winding process approach consisting of an inverted resin impregnation system. The reported impregnation system consists of an inverted design; instead of an impregnating drum, the fibers were run through a bath under specific force followed by mandrel winding. Various parameters such as load viscosity, thickness of tow, impregnation locations, tension, tex, and speed were studied. Furthermore, it was argued that the clean filament winding process could help reduce cleaning solvent usage and facilitate the use of fast-curing resin [21]. Zhou et al. reported the development of a novel sheet winding compression molding process. The process comprises winding the prepregs on the circular mandrel and curing them in the oven using heat-shrink polypropylene film. Furthermore, the produced hollow cylinder was compressed in compression molding. The flat specimens were further analyzed for performance evaluation [22].

4.3 Hybrid manufacturing processes

4.3.1 Injection compression molding (ICM)

ICM is a high-production-rate hybrid manufacturing process that combines a traditional injection molding process followed by a compression molding cycle. Unlike conventional injection molding, where molten charge is injected into a closed mold, in ICM, the polymer melt is fed into an open tool, which is subsequently compressed to produce the final part [23]. The compression stage can be after partial melt filling of the cavity or at the end of the operation sequence [24, 25]. Compared to traditional injection molding, lower molding pressure (~50% lower tonnage) is required in ICM, resulting in a part with higher fiber length retention and higher mechanical properties [23]. In addition, ICM offers other unique advantages, such as reduced residual stress and uneven shrinkage, minimizing molecular orientation, even packing, and reduced density variation [25]. ICM generally makes parts with inserts or details on the part surface that require high dimensional accuracy [23, 25].

4.3.2 Extrusion compression molding (ECM)

ECM is a hybrid manufacturing technique comprising two manufacturing processes (in-line): extrusion and compression molding. A low-shear single-screw extruder is used for the extrusion process. Polymer or composite pellets are gravity- or machine-fed into the extruder through a hopper. The extruder has different zones set up at a temperature higher than the melting temperature of the material being processed. The homogenized viscous molten charge discharged at the end section of the extruder is cut by a hydraulic (or pneumatic) knife at the front of the extruder. The size of the charge depends on the weight of the final part required [23]. The molten charge from the extruder is quickly transferred to a closed mold set in a fast-acting (3–10 inch/min), high-tonnage compression press and compressed to produce a part. Due to the rapid heat loss from the surface of the charge, it tends to cool quickly and will not be able to flow correctly in the mold. So, the transfer and compression of the charge need to be swift before the charge solidifies [23]. Furthermore, molds used for compression molding are often equipped with oil or cartridge heaters to prevent the huge temperature variation from extrusion to the compression process. They are set up at temperatures lower than the processing temperature [23, 26].

Unlike the traditional extrusion process, where a high-shear twin-screw extruder is used to blend composites and produce pellets, a low-shear single-screw extruder used in ECM provides gentle action on the polymer-fiber melt. This leads to less fiber attrition than the twin-screw extruder and injection molding process [23, 27]. Thattaiparthasarathy et al. [28] found that on performing ECM of glass/PP LFT, the average fiber length was reduced from 25 mm to 9.5 mm, with approximately 75% of fibers having a length of more than 5 mm. However, on performing the injection molding of glass/PP LFT, Wang et al. [29] found a significant reduction in the average fiber length of the glass fibers in the composite from 10 mm to 1 mm. Due to the fiber-retaining ability and fast cycle time (0.5–2 min) of the ECM process, it is gaining attention in automotive, military, and infrastructure applications [27, 30]. This process is suitable for making small- to medium-sized semi-structural components [30].

4.3.3 Additive manufacturing -compression molding (AMCM)

The novel AMCM process integrates a high-throughput polymer additive manufacturing process followed by a compression molding process, where the limitations of each process are overcome by the merits of the other. Additive manufacturing can form complex shapes with highly aligned fibers in the deposition direction. However, the part produced has porosity either within the bead or between the layers. On the other hand, parts manufactured via the compression molding process have low porosity but random fiber orientation, depending on the flow progression inside the mold. In AMCM, additively manufactured parts with controlled microstructure are compression-molded to reduce the void/porosity content while maintaining microstructure [31, 32]. The thus-formed composites have comparatively higher mechanical properties than those produced by conventional additive manufacturing or compression molding processes. This process is extensively explained in another chapter of this book.

4.4 Over-molding techniques

Over the past years, over-molding in composites has represented a fascinating and innovative technique in materials engineering. This process combines the strength and rigidity of composite materials with the versatility and customization potential of thermoplastic and thermoset polymers. Over-molding involves the application of a polymer resin onto an existing composite structure, creating a layered or interwoven structure with distinct mechanical properties. This method has gained significant attention in various industries, from aerospace to automotive, due to its ability to enhance product performance, reduce weight, and enable complex part geometries [30, 33–36]. Table 4.1 highlights various over-molding techniques, shedding light on the most common applications.

The following section expands on the benefits and challenges of over-molding techniques.

A few studies will be covered, shedding light on the benefits, challenges, and the evolving landscape of materials science and manufacturing techniques.

R. Akkerman [44] et al. reported an effect on the interfacial strength of thermoplastic composites due to over-molding. The authors studied bonding between the composites,

Table 4.1: Most common over-molding techniques in the industry.

Technique	Description	Application(s)	Reference(s)
Insert over-molding	Preformed components, such as metal or thermoplastic composites, are placed into the mold, and then the polymers are injected around them.	Electronics for encapsulating connectors, automotive for molding metal inserts into plastic components.	[34, 35]
Dual shot over-molding	Injecting the materials into the same mold with two consecutive cycles.	Used for creating multicolor materials, such as soft grips on hard tools.	[37]
Gas-assist over-molding	It uses gas to help hollow out thicker sections of the over-molded part, reducing material usage and enhancing structural integrity.	Medical devices.	[38, 39]
Sheet molding compound (SMC) over-molding	Over-molding unidirectional materials over an SMC intermediate(s) or part(s).	Battery ribs, backrest seat, and various automotive parts.	[40]
Compression over-molding	Over-molding of long fiber thermoplastic composites with a preheated unidirectional material compressed at high-pressure.	Battery enclosures.	[41]
Foam over-molding	Injecting foam into a mold.	Electrical insulation and media-tight electronics.	[42]
Additive manufacturing Compression over-molding	Over-molding of polymer composite onto an additively manufactured metal insert.	High-strength and complex geometry parts.	[31]
In-mold decorating	A graphic overlay is placed into the mold, and the polymer is injected over it.	For decoration and labeled part.	[43]

which is processed through compression molding. First, the intermediates were heated up in a conveyor oven, reaching the melting temperature of the thermoplastic polymer and then pressed in a designed mold at high pressure. Second, the part was injected and over-molded with a semicrystalline polymer such as polyamide 6 (PA6) and PAEK, as shown in Figure 4.4.

The authors evaluated the mechanical properties of the materials by performing tension and shear tests. Insert temperature was used as a variable by considering two manufacturing methods, i.e., Iosipescu inserts and rib-on-plate inserts. For each material, three inserts were over-molded at a temperature below the melting point of the polymers and one sample above the melting temperature. It has been reported that

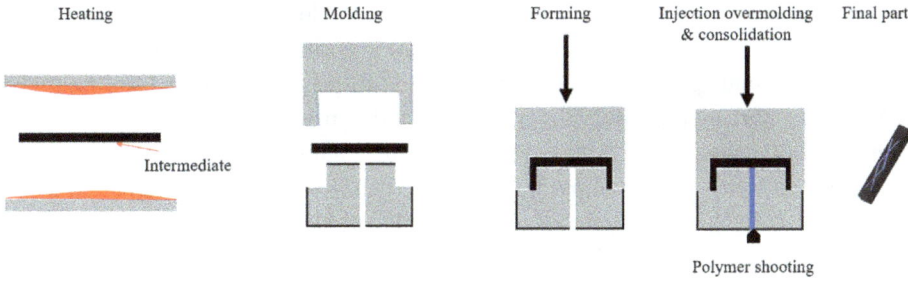

Figure 4.4: Illustration of the injection over-molding process. Glass and carbon fibers were selected as reinforcement; however, PA6 and PAEK were also used as matrices in the intermediates for the injection over-molding.

the tension and shear ultimate strength for the Iosipescu test tremendously increased from 30 MPa at 90 °C to 60 MPa at 270 °C (>melting temperature of polymer). For the rib-on-plate, a significant scattering was noticed in the fully injection-molded specimens, with an increment in the ultimate strength from 15 MPa to 45 MPa in tension. A measure of the interface strength has been introduced based on the degree of crystalline phase melting in semicrystalline polymers, approximated by the maximum interface temperature during over-molding. However, this measure proved conservative at lower interface temperatures but underestimated actual strength at higher temperatures. Higher temperatures led to deformations in the composite laminate due to resin flow, resulting in a nonideal interface structure.

This structural change significantly affected crack initiation and propagation under mechanical stress, indicating that predicting the mechanical performance of over-molded interfaces solely based on idealized interface temperature is insufficient. Instead, a degree of healing for semicrystalline polymers was proposed to define an appropriate processing window. Accurate predictions of processing effects on mechanical performance require further studies tailored to reinforcement structure, part geometry, fiber, resin, and loading conditions.

Alwekar et al. [33] studied the over-molding of long fiber thermoplastics (LFT) with continuous fiber-reinforced thermoplastics (CFRTP) tape using compression over-molding. A charge of LFT, polypropylene glass fiber (PP-GF), was processed using extrusion compression molding (ECM) and was over-molded with CFRTP tape by considering the number of layers as a variable [33, 41].

Bonding interface, fiber alignment, and manufacturing defects were investigated using destructive and nondestructive techniques. On the other hand, the authors reported that the flexural strength and modulus of the over-molded composites were 119% and 77% higher than the LFT composites. The analysis of the interlaminar shear strength tests showed strong bonding at the interface between the CFRTP tape and the LFT substrate; the failure mechanism of the tested samples was monitored using an optical microscope to prove the load transfer between the laminates as well as the excellent fiber dispersion.

The dynamic study of the over-molded part through the low-velocity impact test examined 256% higher energy absorption than the LFT composite. On the nondestructive side, the computed tomography evaluation emphasized the solid interfacial bonding in the over-molded part by reporting less than 1% porosity in both composites. The authors summarized that the compression over-molding techniques could increase the performance and efficiency of the composites with a considerable increase in weight and cost.

Vaidya et al. [40] investigated the over-molding of SMC with continuous reinforcement (woven glass) in manufacturing automotive seat backrests. SMC intermediates were placed on top of the continuous reinforcement in a heated mold (cure temperature of resin) and pressed at high pressure for a dwell time of 5 min to ensure complete consolidation of the seat. Coupons were extracted from different places on the seat to conduct all the mechanical testing. The authors reported an increase of 118% in flexural strength and 89% in stiffness in the SMC over-molded part. The scanning electron microscopy images evaluated for the ILSS samples showed sufficient bonding between the SMC and the SMC-OM despite the nonsignificant improvement in the ILSS strength, highlighting that one layer of woven glass was embedded in the seat. The SMC over-molded part performed better than the SMC with an enhancement of 15–20%, as they mentioned. Vibration tests emphasized the damping ratio and stiffness improvement in the SMC over-molded part. The findings of this study demonstrate that SMC over-molding, with a small amount of continuous reinforcement, offers the ability to customize mechanical properties and facilitates the production of nearly net-shaped components. Through the use of micro-X-ray computed tomography, it was confirmed that the connection between the sheet molding compound and the over-molding insert exhibited significant strength. This method presents an economically viable means to produce high-volume, cost-effective fiber-reinforced composite components suitable for use in the automotive sector.

The utilization of over-molding in composite manufacturing offers a multitude of compelling advantages. Over-molding allows for seamless integration of distinct materials, enabling the tailoring of mechanical properties and the creation of components with enhanced structural performance. This versatility is invaluable in industries demanding precise material characteristics. Moreover, over-molding empowers the production of intricate and complex part geometries, expanding the realm of design possibilities. Many studies elaborate on its efficacy in achieving intricate forms that would be challenging or unattainable through conventional manufacturing methods.

4.5 Fiber-reinforced metal laminates

Fiber-reinforced metal laminates, popularly abbreviated as FML, are low-cost, lightweight, and high-strength materials used in various applications. Combining fiber-reinforced polymer composites with metals bestows the advantages of both metals and advanced composites. The development of FML began addressing the search for lightweight materials in aerospace structures [45]. The prime requirements were a combination of high specific strength, high elastic modulus, toughness, corrosion resistance, and fatigue resistance. Fiber-reinforced composites fulfill these demands except for fracture toughness. High-strength aramid fibers were introduced to address the fatigue resistance, followed by the development of aramid fiber-reinforced aluminum laminate (ARALL) [46]. Figure 4.5 depicts the classification of FMLs.

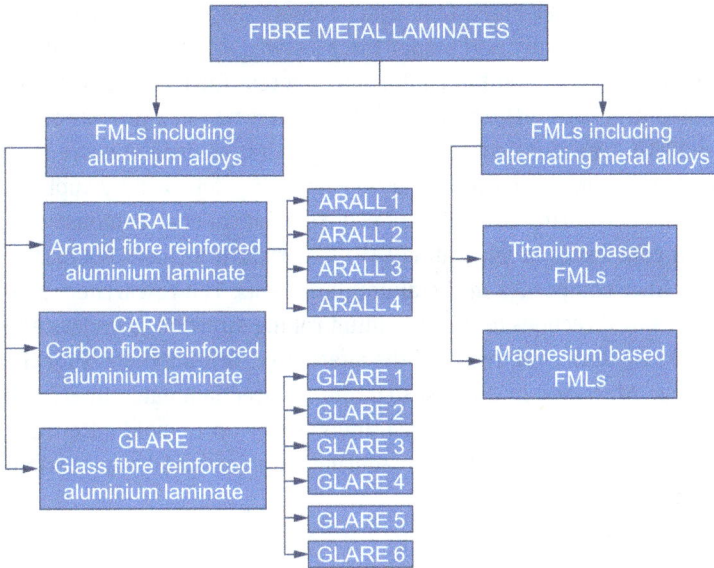

Figure 4.5: FML classification of fiber-reinforced metal laminates (reproduced with permission from Elsevier, copyright © 2021).

For ARALL, different grades of aluminum were combined with aramid fibers, whereas for glass-reinforced aluminum laminate (GLARE), the grade of aluminum and the orientation of unidirectional glass reinforcement are at variable angles like 0/0, 90/90, 0/90, etc. For CARALL, i.e., carbon fiber-reinforced metal laminate, the material was developed by replacing aramid reinforcement with carbon fiber/epoxy prepregs. FML has multiple advantages, such as high strength, fracture toughness, high fatigue resistance, high impact strength, low density, durability, and cost-saving [47].

4.5.1 Forming of FMLs

FML is usually processed by compression molding-assisted stamping. In this method, the stack of aramid/carbon or glass laminates is placed in the required orientation, and aluminum skin is placed on it. The whole laminate can be heated in an oven or compression molding press, followed by curing and cooling. The development of manufacturing technology gained traction in the last two decades. Edwardson et al. reported the laser-assisted forming process for the FML material, and it was noted that the utility of laser-assisted FML forming helps with complex shapes and bends [48]. Gisario et al. used a laser forming process to fabricate FML with specific curvature. It was reported that laser forming of FML allows the forming of curved parts with accuracy and quality [49]. Kalyanasundaram et al. studied the effect of process parameters during the forming process of self-reinforced PP-based FML using a dome shape. In this configuration, it was found that the temperature and binder force showed an effect on the major strain [50]. Behrens et al. demonstrated the conforming process for FML consisting of forming FML in one-step processes [51]. Zhang et al. explored a two-step process to develop aluminum-based FML. The study comprises two steps: first, shaping and strengthening the aluminum sheet by heat quenching and thermal aging, followed by applying epoxy glass fiber laminate and curing (second step) [52]. Yelamanchi et al. demonstrated the utility of 3D-printed composites in fabricating FML materials. It includes hot stamping of 3D-printed glass fiber composites in the aluminum sheets. The researchers analyzed the 0° and 45° printing directions to the aluminum rolling direction to understand the effect of orientation [53]. Roth et al. reported the alternative forming process for the FMLs, including two-stage processes. In the first step, the uncured laminates, interfacial elastomers, and the aluminum sheet were shaped, followed by an in-mold cure using the required temperature. It was argued that the alternative process bestows a certain degree of formability and shortens cycle time [54].

6 Conclusion

This book chapter comprises a systematic review of conventional composite manufacturing techniques and hybrid manufacturing techniques. It includes conventional techniques such as compression molding, injection molding, braiding, commingling, automated tape placement, and filament winding. It comprises features of techniques and dependent evaluation of performance properties, such as fiber orientation, changes in mechanical and thermal properties, etc. Furthermore, hybrid manufacturing techniques include injection compression molding (ICM), extrusion compression molding, and additive manufacturing-compression molding (AMCM). These hybrid techniques help reduce cycle time, tooling, and operating costs, preserve performance properties by imparting crystallinity, and maintain reinforcement orientation and fiber length. Over-molding is another hybrid technique, usually combined with injection and compression molding. It usually includes the addition of metal inserts, carbon fibers, and its laminate over-molding on short fiber and long fiber composites, which results in excellent overall mechanical properties. Moreover, the chapter discusses fiber-reinforced metal laminates (FMLs), including their properties, hybrid combinations, and forming techniques.

Acknowledgment: The authors acknowledge the support from the US Department of Energy (DOE), Office of Energy Efficiency and Renewable Energy, and Advanced Materials and Manufacturing Office. This manuscript has been authored by UT-Battelle, LLC, under contract DE-AC05-00OR22725 with the US Department of Energy (DOE). The US government retains and the publisher, by accepting the article for publication, acknowledges that the US government retains a nonexclusive, paid-up, irrevocable, worldwide license to publish or reproduce the published form of this manuscript, or allow others to do so, for US government purposes. DOE will provide public access to these results of federally sponsored research in accordance with the DOE Public Access Plan (http://energy.gov/downloads/doe-public-access-plan).

References

[1]   U. Marathe, M. Padhan, J. Bijwe Exploration of pros and cons of polyetherimide solutions with varying concentrations as the sizing agents for graphite fibers in graphite fabric-PAEK composites, Journal of Materials Research and Technology, 2021, 14: 2085–2095.
[2]   J.A. Brydson Plastics Materials, Elsevier, Oxford, 1999.
[3]   P.K. Mallick Fiber-reinforced Composites: Materials, Manufacturing, and Design, CRC press, Boca Raton, FL, 2007.
[4]   J. Brydson Principles of the processing of plastics, in: Plastics Materials, 7th ed, Butterworth-Heinemann, Oxford, UK, 1999, pp. 158–182.
[5]   N.M. Meliande et al. Ballistic properties of curaua-aramid laminated hybrid composites for military helmet, Journal of Materials Research and Technology, 2023, 25: 3943–3956.
[6]   H. Jiang et al. High-efficient improvement in flexural properties of carbon/Kevlar-fiber hybrid composites by CNT-toughening only between xenogeneic fiber-layers, Thin-Walled Structures, 2023, 190: 110984.
[7]   A. Proietti et al. Thermo-formable hybrid carbon fibre laminates by composite recycling, The International Journal of Advanced Manufacturing Technology, 2023, 7(127): 1–13.
[8]   S. Doagou-Rad et al. Investigation of conductive hybrid polymer composites reinforced with copper micro fibers and carbon nanotubes produced by injection molding, Materials Today Communications, 2019, 20: 100566.
[9]   Y. Zhang et al. Direct injection molding and mechanical properties of high strength steel/composite hybrids, Composite Structures, 2019, 210: 70–81.
[10] G.B. Carvalho, S.V. Canevarolo Jr, J.A. Sousa Influence of interfacial interactions on the mechanical behavior of hybrid composites of polypropylene/short glass fibers/hollow glass beads, Polymer Testing, 2020, 85: 106418.
[11] S.K. Lee et al. Mechanical properties of PP/glass fiber/kenaf/bamboo fiber-reinforced hybrid composite, Fibers and Polymers, 2021, 22: 1460–1465.
[12] Y. Bai et al. Highly thermally conductive yet electrically insulative polycarbonate composites with oriented hybrid networks assisted by high shear injection molding, Macromolecular Materials and Engineering, 2022, 307(1): 2100632.
[13] G.V. Marrivada et al. effect of addition of graphene nanoplatelets on the mechanical properties of triaxially braided composites, Advanced Composite Materials, 2023, 32(2): 182–210.

[14] Q. Gu et al. fabrication and braiding angle effect on the improved interlaminar shear performances of 3D braided sandwich hybrid composites, Journal of Materials Research and Technology, 2023, 25: 5795–5806.

[15] A. Abdkader et al. Development of an innovative glass/stainless steel/polyamide commingled yarn for fiber–metal hybrid composites, Materials, 2023, 16(4): 1668.

[16] H. Awais et al. effect of fabric architecture on the shear and impact properties of natural fibre reinforced composites, Composites Part B: Engineering, 2020, 195: 108069.

[17] R.M.D. Benedetto et al. Development of hybrid steel-commingled composites CF/PEEK/BwM by filament winding and thermoforming, COMPOSITES SCIENCE AND TECHNOLOGY, 2022, 218: 109174.

[18] Z. Qureshi et al. In situ consolidation of thermoplastic prepreg tape using automated tape placement technology: Potential and possibilities, Composites Part B: Engineering, 2014, 66: 255–267.

[19] C. Stokes-Griffin, P. Compston A combined optical-thermal model for near-infrared laser heating of thermoplastic composites in an automated tape placement process, Composites Part A: Applied Science and Manufacturing, 2015, 75: 104–115.

[20] M. Azeem et al. Application of filament winding technology in composite pressure vessels and challenges: A review, Journal of Energy Storage, 2022, 49: 103468.

[21] S. Pandita et al. Clean wet-filament winding–Part 1: Design concept and simulations, Journal of Composite Materials, 2013, 47(3): 379–390.

[22] Y. Zhou et al. Experimental investigation of thermoset composite laminates manufactured with a novel sheet-winding compression-molding (SWCM) process, Materials Today Communications, 2023, 35: 106034.

[23] U. Vaidya Composites for Automotive, Truck and Mass Transit: Materials, Design, Manufacturing, DEStech Publications, Inc, Lancaster, PA, 2011.

[24] S.C. Chen, Y.C. Chen, H.S. Peng Simulation of injection–compression-molding process. II. Influence of process characteristics on part shrinkage, Journal of Applied Polymer Science, 2000, 75(13): 1640–1654.

[25] D. Loaldi et al. Manufacturing signatures of injection molding and injection compression molding for micro-structured polymer fresnel lens production, Micromachines, 2018, 9(12): 653.

[26] A. Walkare et al. design and analysis of extrusion-compression molding setup for processing of glass fiber reinforced polypropylene composites, Materials Today: Proceedings, 2023, 72: 3017–3022.

[27] H. Ning et al. A review of Long fibre thermoplastic (LFT) composites, International Materials Reviews, 2020, 65(3): 164–188.

[28] K.B. Thattaiparthasarathy et al. Process simulation, design and manufacturing of a long fiber thermoplastic composite for mass transit application, Composites Part A: Applied Science and Manufacturing, 2008, 39(9): 1512–1521.

[29] J. Wang et al. Shear induced fiber orientation, fiber breakage and matrix molecular orientation in long glass fiber reinforced polypropylene composites, Materials Science and Engineering: A, 2011, 528(7–8): 3169–3176.

[30] J. Andrzejewski, M. Szostak Preparation of hybrid poly (lactic acid)/flax composites by the insert overmolding process: Evaluation of mechanical performance and thermomechanical properties, Journal of Applied Polymer Science, 2021, 138(2): 49646.

[31] D.K. Pokkalla et al. A novel additive manufacturing compression overmolding process for hybrid metal polymer composite structures, Additive Manufacturing Letters, 2023, 5: 100128.

[32] D.K. Pokkalla et al. Anisotropic mechanical properties of polymer composites from a hybrid additive manufacturing-compression molding process using X-ray computer tomography, in: Nondestructive Characterization and Monitoring of Advanced Materials, Aerospace, Civil Infrastructure, and Transportation XVI, SPIE, 2022.

[33] S. Alwekar et al. manufacturing and characterization of continuous fiber-reinforced thermoplastic tape overmolded long fiber thermoplastic, Composites Part B: Engineering, 2021, 207: 108597.

[34] M. Bakr et al. Effect of overmolding process on the integrity of electronic circuits, in: 2019 22nd European Microelectronics and Packaging Conference & Exhibition (EMPC), IEEE, 2019.

[35] M. Grujicic Injection overmolding of polymer–metal hybrid structures, in: Joining of Polymer-Metal Hybrid Structures: Principles and Applications, 2018, pp. 277–305.

[36] S. Wasti et al. Bamboo fiber overmolding textile grade carbon fiber tape and bamboo fiber polypropylene composites, Sampe Journal, 2023, 59(2): 22–29.

[37] J. Markarian Overmolding continues to grow & find creative new applications: New materials and processes are helping to expand how overmolding can be used, Plastics Engineering, 2017, 73(5): 24–27.

[38] M. Hansen Overmolding: A multifaceted medical device technology, Medical Device and Diagnostic Industry, Jan. 2006, 2006.

[39] A.M. Hosseini et al. The effects of gas assisted injection molding on the mechanical properties of medical grade thermoplastic elastomers, Polymer Testing, 2014, 38: 1–6.

[40] U. Vaidya et al. Manufacturing demonstration of automotive seat backrest using sheet molding compound and overmolding with continuous reinforcement, Applied Composite Materials, 2022, 29(3): 1367–1391.

[41] S.P. Alwekar Innovations in Aligned and Overmolded Long Fiber Thermoplastic Composites, PhD thesis from University of Tennessee Knoxville, Knoxville, TN, 2021.

[42] C. Ott, D. Drummer Low-stress over-molding of media-tight electronics using thermoplastic foam injection molding, Polymer Engineering and Science, 2021, 61(5): 1518–1528.

[43] H. Kao In-Mold Decorating: A Review of Process and Technology, Plastics Engineering, 2018, 74(7): 40–43.

[44] R. Akkerman, M. Bouwman, S. Wijskamp Analysis of the thermoplastic composite overmolding process: Interface strength, Frontiers in Materials, 2020, 7: 27.

[45] J. Remmers, R. De Borst Delamination buckling of fibre–metal laminates, Composites Science and Technology, 2001, 61(15): 2207–2213.

[46] G.R. Villanueva, W. Cantwell The high velocity impact response of composite and FML-reinforced sandwich structures, Composites Science and Technology, 2004, 64(1): 35–54.

[47] U. Marathe et al. Processing of PAEK-graphite fabric composites–Pros and cons of film technique over powder sprinkling technique, Composites Part B: Engineering, 2021, 215: 108804.

[48] S. Edwardson et al. laser forming of fibre metal laminates, Lasers in Engineering, 2005, 15(3): 233–255.

[49] A. Gisario, M. Barletta Laser forming of glass laminate aluminium reinforced epoxy (GLARE): On the role of mechanical, physical and chemical interactions in the multi-layers material, Optics and Lasers in Engineering, 2018, 110: 364–376.

[50] S. Kalyanasundaram et al. effect of process parameters during forming of self reinforced–PP based Fiber Metal Laminate, Composite Structures, 2013, 97: 332–337.

[51] B. Bernd-Arno et al. Forming and joining of carbon-fiber-reinforced thermoplastics and sheet metal in one step, Procedia Engineering, 2017, 183: 227–232.

[52] Q. Zhang et al. The experimental study on the mechanical properties of fiber-reinforced metal laminates using an innovative heat-solid integrated forming technology, Metals, 2023, 13(7): 1199.

[53] B. Yelamanchi et al. The fracture properties of fiber metal laminates based on a 3D printed glass fiber composite, Journal of Thermoplastic Composite Materials, 2023, 36(2): 815–835.

[54] S. Roth et al. A new process route for the manufacturing of highly formed fiber-metal-laminates with elastomer interlayers (FMEL), The International Journal of Advanced Manufacturing Technology, 2019, 104: 1293–1301.

Kazi Md Masum Billah

5 Hybrid additive manufacturing for composites

5.1 Introduction

Additive manufacturing (AM), commonly known as 3D printing is the process of depositing and joining materials in a layer-upon-layer fashion to make parts from 3D computer-aided design (CAD) model data. In general, there are seven different categories of AM processes available such as (a) material extrusion, (b) binder jetting, (c) powder bed fusion, (d) VAT photopolymerization, (e) direct energy deposition, (f) material jetting, and (g) sheet lamination. The unique characteristics among all seven process categories of AM are complex design part fabrication, exploration of new materials, and composite manufacturing. Feedstock for each AM process category is different and well-bounded by the definition, working principles, and their respective machine instrumentation. AM processes are relatively new and became available in commercial sectors in the late 1990s. Many engineering applications have been identified to date, and a continuous thrust is still being made to mature these technologies by instrumenting with new tools and technologies. The common goal of each AM process is to explore new material capabilities, such as composite structure fabrications. While traditional manufacturing, such as subtractive and forming processes, has limitations on the manufacturability of specific composites, AM offers benefits in the composite manufacturing community to develop new classes of composite materials and structures. This chapter will elucidate the effort of hybrid composite manufacturing by combining state-of-the-art AM machines with innovative tools and devices. Hybrid additive manufacturing (HyAM) is an advanced manufacturing process that combines multiple tools and technologies to fabricate composite yet monolithic structures. Often, HyAM is accomplished by adding tools to the existing additive manufacturing (AM) machine to deposit discrete materials. With the intention of adding multiple materials during the fabrication process, HyAM typically integrates various tools using a computer-integrated manufacturing system. Several existing HyAM capabilities are

Acknowledgment: This work was supported in part by the U.S. Department of Energy, Office of Science, and Office of Workforce Development for Teachers and Scientists (WDTS) under the Visiting Faculty Program-Student program. Additionally, this material is based upon work supported by the National Science Foundation under Grant No. 2301925. The views and conclusions contained herein are those of the author and should not be interpreted as necessarily representing the official policies or endorsements, either expressed or implied, of ORNL or the U.S. Government.

Kazi Md Masum Billah, Mechanical Engineering, University of Houston-Clear Lake, e-mail: billah@uhcl.edu

https://doi.org/10.1515/9783111019543-005

available across the AM process categories; however, this chapter will focus on the material extrusion-based AM process used to fabricate hybrid composites and their structures.

Material extrusion AM process, particularly fused filament fabrication (FFF) is one of the most commonly used technologies among the seven process categories. The main reasons for the large-scale presence of FFF in the market are the low cost of machines, the versatility of thermoplastics-based filaments, and the ease of operations. The evolution of the FFF system led to the development of desktop-sized (typically USD 200) to large-scale machines (more than USD 500,000). The desktop and production-grade machines are primarily developed for creative design engineers and manufacturing shops to aid the prototyping and rapid replacement of functional part fabrication using neat thermoplastics. On the other hand, large-scale machines are developed for mold and tooling applications. For example, Oak Ridge National Laboratory (ORNL) developed the big area additive manufacturing (BAAM) machine, which uses thermoplastic pellets as feedstock [1]. The typical deposition rate of the BAAM machine is 100 lb. per hour using a single-screw extruder, and the build volume is 10 ft.×8 ft.×6 ft.

The increasing demand for 3D-printed parts with enhanced mechanical properties and multifunctionality has led to the development of hybrid machines. A typical hybrid FFF-based AM machine can fabricate a composite part using two different materials: thermoplastic ABS and thermoplastic PLA. Depending on the application, thermally and electrically conductive or resistive materials such as copper wire, nichrome wires, carbon fiber, etc., can be deposited on a 3D-printed substrate by augmenting the deposition tool, thus fabricating hybrid composites and their structures.

Figure 5.1: Steps in manufacturing 3D RF circuit components [1]: (a) PLA substrate transmission line and (b) PLA substrate annular ring resonator with a copper sheet base and subminiature connectors attached to the microstrip annular ring resonator.

Composite structures with complex and conformal architecture that consist of multiple materials offer great benefits such as a high strength-to-weight ratio, topologically optimized shape, and embedded multifunctionality for many applications, including aerospace, automotive, industrial tooling, biomedical, and defense. An example of a hybrid composite structure of a 3D-printed antenna using thermoplastic PLA and conductive copper is shown in Figure 5.1. The following section discusses the state-of-the-art FFF-based HyAM of hybrid composites.

5.2 Hybrid composites using HyAM

A typical FFF system consists of a dual extrusion system depositing two materials. The purpose of the dual extrusion system is to deposit multiple materials during the fabrication process and develop support structures for overhang parts. Hybrid composites are produced in many forms from this simple concept of multi-material deposition using a dual extrusion system. An example of a dual extrusion FFF system that deposits two different thermoplastic materials to fabricate composite structures is shown in Figure 5.2.

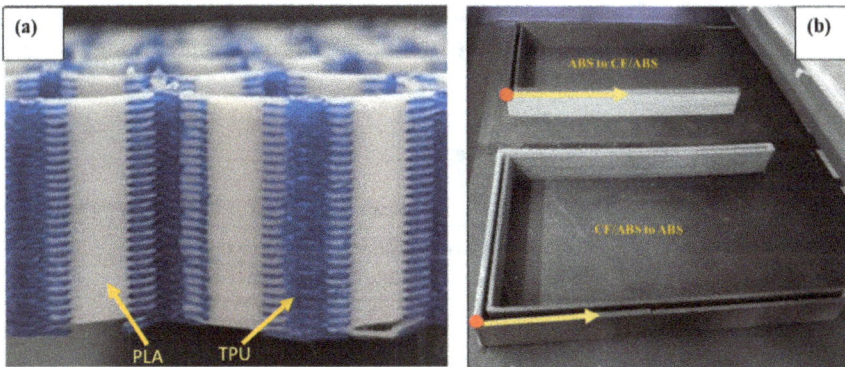

Figure 5.2: (a) Representative dual extruder 3D printer for hybrid composite structures using neat thermoplastic PLA and TPU [2]. (b) Representative dual material 3D printing for neat ABS and ABS with chopped carbon fiber for a gradient structure [3].

5.3 Materials for HyAM-based composites

Primarily, the FFF system uses thermoplastic feedstock that comes as a filament in a spool. The most common feedstocks for FFF systems are polylactic acid (PLA), acrylonitrile butadiene styrene (ABS), polyethylene terephthalate glycol (PETG), thermoplastic polyurethane (TPU), polyamide (Nylon), acrylonitrile styrene acrylate (ASA), polycarbonate (PC), polyether ether ketone (PEEK), polyetherimide (ULTEM), and polyvinyl alcohol (PVA). Hybrid composites are manufactured by combining these materials during the 3D printing processes. As shown in Figure 5.3, the FFF multi-material hybrid structure is manufactured using thermoplastic ABS and PC.

Figure 5.3: Hybrid composite structures are manufactured in an FFF machine using colored (white) ABS and (red) PLA.

Considering the lower mechanical strength of the FFF thermoplastic part compared to the injection-molded counterpart, both short and continuous fiber-based materials are used. A detailed list of the materials that are being used in HyAM-based composite manufacturing by adding thermoplastic feedstock before printing and during printing is listed in Table 5.1:

Table 5.1: List of materials used for hybrid composite manufacturing using thermoplastic feedstocks in the FFF system.

Fibers	Mixed with thermoplastic feedstock before printing?	In-situ mixing with thermoplastic?
Short carbon fiber (SCF)	Yes	No
Short glass fiber (SGF)	Yes	No
Short wood fiber (SWF)	Yes	No
Continuous carbon fiber (CCF)	No	Yes
Continuous glass fiber (CGF)	No	Yes
Continuous Kevlar fiber (CKF)	No	Yes
Continuous copper wire (CCW)	No	Yes
Continuous nichrome wire (CNW)	No	Yes

Continuous wires, such as copper wire and nichrome wire, are used to fabricate conductive and resistive circuitry-based hybrid composites, respectively.

5.4 Methods and applications

Fabricating hybrid composites using FFF machines involves a series of methods and techniques designed to integrate reinforcing materials within a thermoplastic matrix effectively. These methods are essential to harness the benefits of both materials, resulting in composites with enhanced mechanical, thermal, and electrical properties. This section explores the primary methods of producing hybrid composites from small-scale to large-scale machines using FFF technology.

5.4.1 Hybrid composites using small-scale FFF machines

Several methods encompass the FFF system to fabricate fiber-reinforced hybrid composites. Figure 5.4 shows three different FFF-based hybrid composite manufacturing systems available on the market. All three configurations commonly use thermoplastics. In the case of the: (a) traditional dual extrusion FFF system, different thermoplastics and chopped fiber-reinforced composite filaments are used; (b) in-situ fiber impregnation and fusion system, continuous fibers are used; and (c) pre-impregnated fiber extrusion system, thermoplastic resin-infused continuous fiber-based feedstocks are used. Augmentation of the existing single extruder FFF machine with a dual extruder enables the manufacturing of multi-material hybrid composites. This is one of the most common methods for fabricating hybrid composites. This technique utilizes FFF machines equipped with two extruders: one for the thermoplastic matrix and the other for the reinforcing material, as shown in Figure 5.4 (a). The dual extrusion method allows for precise control over the deposition of both materials, creating complex geometries and tailored material properties.

A typical desktop-scale dual extrusion process involves material preparation, deposition, and welding of deposited beads. In the material preparation phase, thermoplastic filaments such as ABS, PLA, PC, PETG, ULTEM, PEEK, and reinforcing fibers (e.g., carbon, glass, or Kevlar) are prepared and loaded into separate extruders. In dual extrusion systems, the recent addition of continuous fiber-based hybrid composites has been received well for functional prototyping of composite structures in the manufacturing community. In the layer-by-layer deposition phase, the FFF machine deposits the thermoplastic filament to form the matrix layer, followed by the reinforcing fiber in specified regions to enhance strength and stiffness. During the slicing of the digital part, such as the 3D CAD part, materials A and B can be assigned in the slicing software. Generated tool paths with material-specific processing parameters are fed into the 3D printer to fabricate bi-material composite parts. After deposition, the welding of the beads ensures structural shape retention. Optimized extrusion temperatures and print speeds warrant strong adhesion between the thermoplastic matrix and the fibers, resulting in a cohesive hybrid composite structure.

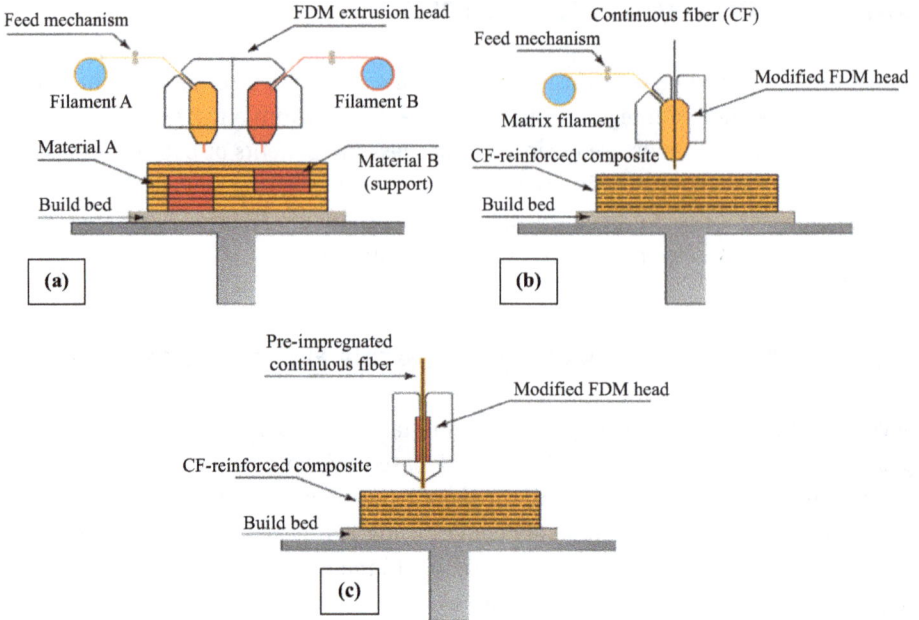

Figure 5.4: FFF-based hybrid composite manufacturing technologies: a) traditional dual extrusion FFF, b) in-situ fusion of carbon fibers with molten thermoplastic in the nozzle, and c) extrusion of pre-impregnated fibers [4].

Thermoplastic matrix-infused and continuous fiber-reinforced composites can be manufactured using a modified single extruder 3D printer, as shown in Figures 5.4 (b) and (c). In this case, fiber bundles are fed through a separate path; however, fibers are allowed to mix with molten thermoplastic filament within a modified hot-end extruder and co-extrude at the nozzle exit. For example, a single extruder 3D printing nozzle was developed to deposit continuous fiber-reinforced thermoplastic-based hybrid composite in an FFF machine [4]. Polylactic acid (PLA) was used as the matrix, with carbon or natural jute fibers as reinforcements. The developed extruder impregnated the fiber bundles before deposition, as shown in Figure 5.5 (a). As shown in Figure 5.5 (b), the unidirectional jute fibers created plant-sourced composites, while carbon fibers yielded superior mechanical properties compared to both jute-reinforced and unreinforced thermoplastics. The demonstration of the hybrid composite 3D printing and fabricated parts are shown in Figures 5.5 (c) and (d).

Similar to the dual extrusion system, tool path planning can be generated in slicing software. In addition to dual extrusion and in-situ fusion, continuous fiber-based hybrid composites can be fabricated by preparing the fibers. As an example of carbon fiber (CF)-based composites, prepreg CF strands can be fed through a single extruder. Various sizing materials and prepreg resins are used with continuous CF strands to fabricate tailored composites. As shown in Figures 5.6 (a) and (b), a single extruder 3D

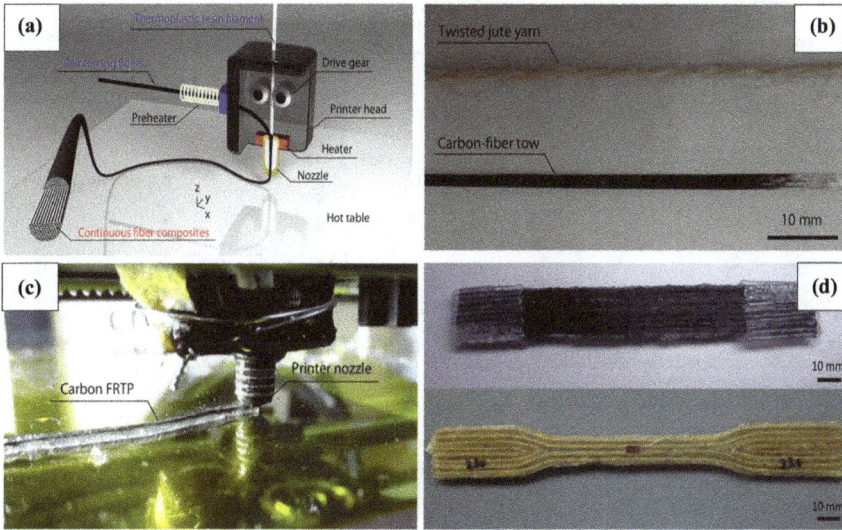

Figure 5.5: (a) Schematic of the 3D printer head to produce continuous fiber-reinforced thermoplastics using in-nozzle impregnation based on fused filament fabrication. (b) Continuous fiber reinforcements are used for 3D Printing. (c) Photograph of the 3D printing of a continuous carbon fiber- reinforced thermoplastic hybrid composite. (d) 3D-printed carbon fiber-reinforced thermoplastic (top) and jute yarn-reinforced thermoplastic composite (bottom) [5].

printer nozzle was developed to manufacture hybrid composites [6]. The pin-assisted fiber impregnation allowed the fiber bundle to be consolidated in one strand, thus improving the bonding between fibers and thermoplastic matrix materials. The developed machine tools showed that the mechanical properties are tailored and improved compared to the commercially available 3D printing machine, which is characterized by fiber pull-out testing (Figures 5.5 (c) to (e)).

5.4.1.1 Advantages and disadvantages of small-scale 3D-printed hybrid composites

The incorporation of continuous fibers significantly increases the strength, stiffness, and impact resistance of the printed parts. The ability to tailor fiber placement and orientation allows customized reinforcement, optimizing the performance for specific applications. The AM process enables complex geometries and integrated structures that are challenging to achieve with traditional manufacturing methods. Despite the benefits of complex geometry-based fabrication in AM, certain limitations are inherent in AM-based composites. While it is possible to tailor the fiber placement and modulate the fiber volume fraction and fiber wetting, it is a well-known challenge that the AM process cannot fabricate a void-free part. For instance, compression molding-based composites are void-free and defect-free parts. In contrast, AM-based composites have meso-

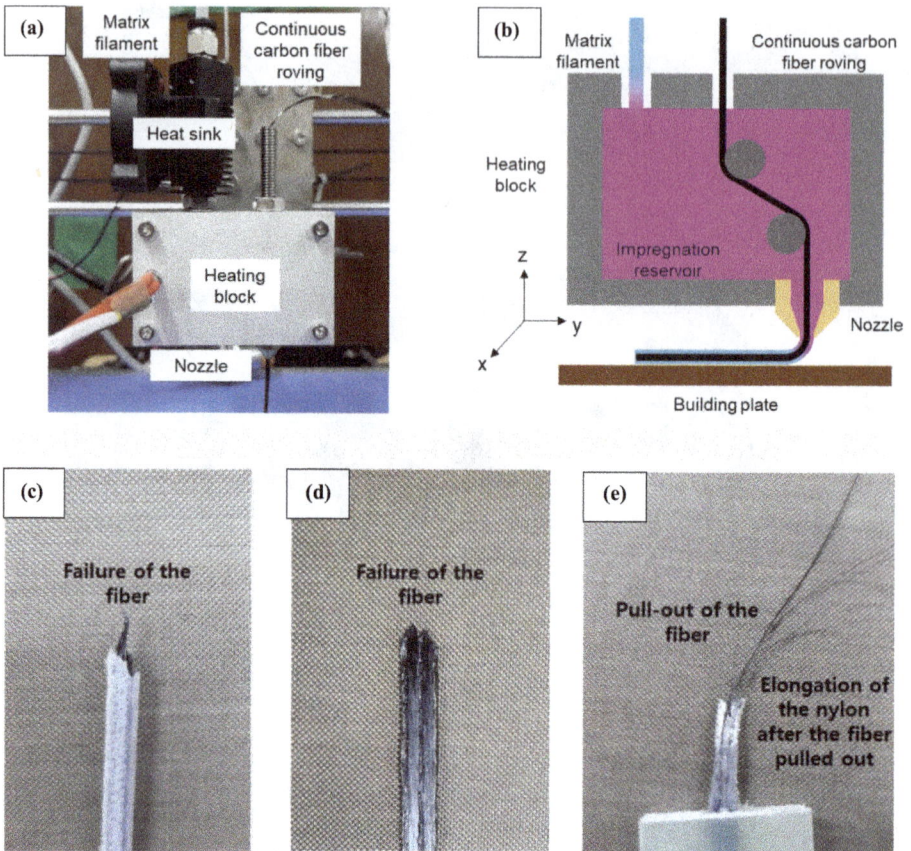

Figure 5.6: Modified single extruder 3D-printing system for continuous carbon fiber-reinforced thermoplastics-based hybrid composite manufacturing: (a) actual printing system, (b) schematic illustration showing the in-situ pin-assisted melt impregnation printing head, (c) commercially developed Mark Two 3D-printed hybrid composite tested for mechanical characterization, (d) in-nozzle impregnation 3D printer, and (e) in-situ pin-assisted melt impregnation 3D printer equipped with four pins [6].

structure porosity within the beads and microstructure voids between the fiber and matrix interface. These voids possess weaker mechanical properties for fatigue and creep-based applications.

5.4.1.2 Applications of hybrid composites fabricated using 3D printing

Direct applications of 3D-printed hybrid composites, such as fiber-reinforced thermoplastics, are emerging in the manufacturing community for tooling applications. For example, many applications for automotive vehicles have been identified for high-performance composite parts, such as seat panels, electrical enclosures, and interior body panels. In the case of aerospace applications, there is strong potential for lightweight and strong components for aircraft and spacecraft; however, significant characterization and testing are still required to qualify for in-flight applications. Reinforced structures are also ideal candidates for sporting goods. HyAM-based products are attractive due to their human-specific design and on-demand manufacturability. Customized and durable medical implants and prosthetics are examples of biomedical applications of hybrid composites. Despite the strong potential of on-demand and mass customization manufacturing with tailored properties for a wide range of applications, there is a significant need for research and technology development for end-use and functional parts of HyAM-based composites.

5.4.2 Hybrid composites by adding tools in the scalable FFF system

Hybrid composites that consist of 3D-printed thermoplastics and conductive or resistive elements can be manufactured by developing customized tools for the FFF system. Two different types of tools are used to fabricate these hybrid composites and structures: (a) co-extrusion and (b) embedding tool.

5.4.2.1 Co-extrusion-based hybrid composite 3D printing

Co-extrusion-based hybrid composites are fabricated by adopting wire co-extrusion technologies. Co-extrusion-based hybrid composites are an innovative approach that involves the simultaneous extrusion of wires with thermoplastic materials. This technology allows the creation of advanced composite structures with integrated functionalities, such as electrical conductivity and resistance. Both electrically conductive and resistive wires can be co-extruded with thermoplastic feedstock. As an example of wire co-extrusion-based 3D printing in a BAAM machine, ORNL demonstrated the 3D printing of nichrome wire co-extruded composite molds, as shown in Figure 5.7.

The general working principle of this wire co-extrusion 3D printing for hybrid composite structures consists of materials preparation, extrusion system design, tool path generation, 3D printing, and post-processing. Co-extrusion feedstocks such as wires and thermoplastic composites are prepared in the material preparation phase. Electrically conductive and resistive wires, such as copper and nichrome wires, can be fed through the co-extrusion nozzle, as shown in Figure 5.7. The modified co-extrusion system allows the 3D printing of wires and thermoplastic resins simultaneously. The extruder is equipped with additional feeding mechanisms for both the conductive and resistive wires. The nozzle is designed to co-extrude the wires along with the thermoplastic material, ensuring proper embedding and alignment within the printed structure. To precisely control the wire and resin feeding and deposition rate, custom-developed slicing

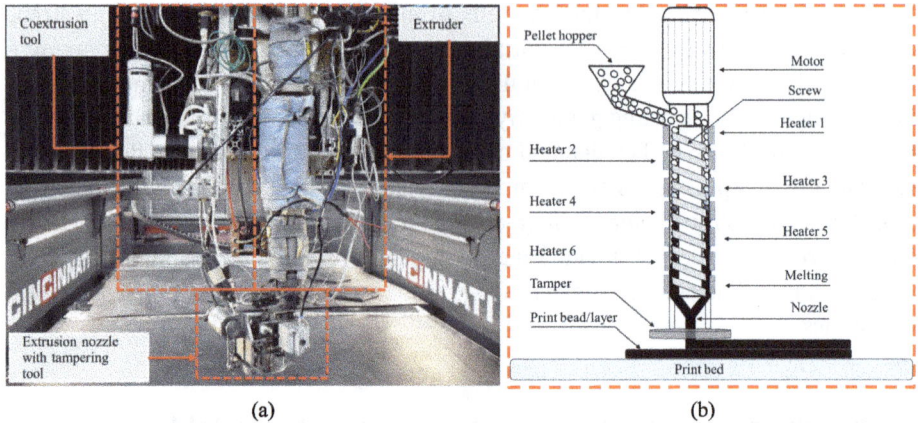

Figure 5.7: (a) Integrated wire co-extrusion tool and extruder mounted on the BAAM gantry developed at ORNL and (b) schematic of the BAAM single-screw extruder with wire co-extrusion system for composite mold 3D printing [7].

software is used to generate tool paths that incorporate both types of wires. The ORNL slicing software allows precise control over the placement, orientation, and length of the wires, depending on the desired functionality of the printed part [8]. The print head follows the programmed tool paths, depositing the composite material layer by layer. Each layer is aligned and bonded/welded with the previous layer, forming a continuous and functional hybrid composite structure.

For large-scale part fabrication, the typical post-processing operation is the surface finishing facing operation in a milling machine. The facing operation is performed on the part in a controlled manner to retain dimensional accuracy and mechanical integrity. Depending on the application, additional post-processing steps, such as electrical testing and insulation, may be required to achieve the desired functionality and finish.

5.4.2.2 Applications of wire co-extrusion hybrid composites

The embedded electrical circuits are a prime example of wire co-extrusion-based hybrid composites: embedding metallic wires within thermoplastic structures can create integrated electrical circuits for various electronic applications. Compared to conductive ink-based electrical circuits, metallic wire has higher conductivity and eliminates the need for a sintering process. Therefore, higher structural integrity is obtained compared to ink-based circuitry. Wire-embedded mold is another example of co-extrusion technology for self-heating out-of-oven autoclave molds. Nichrome wires can be embedded to create resistive heating elements for heating purposes instead of conductive wires. As shown in Figure 5.8, a self-heating mold was fabricated at ORNL BAAM with co-extrusion technology. Uniform temperature distribution during the thermal testing ensured the structural integrity of the mold for the composite tooling application.

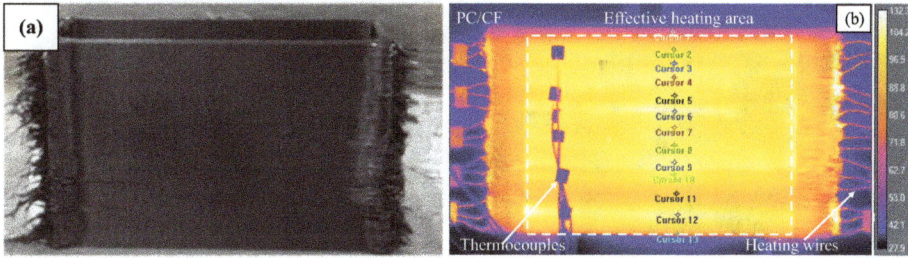

Figure 5.8: (a) Nichrome wire embedded self-heating mold fabricated in ORNL BAAM with co-extrusion technology and (b) thermal testing of self-heating mold that was prepared for testing after post-processing machining operation [9].

In addition to the embedded electronics and self-heating mold tooling application, the combination of conductive and resistive wires within a single thermoplastic matrix allows the creation of smart structures with integrated sensing, heating, and conductive pathways. This embedded sensing capability will enable smart tooling for Industry 4.0 applications.

5.4.2.3 Wire embedding tool-based hybrid composite 3D Printing

Thermal wire embedding is another method of depositing conductive wires within 3D-printed parts. In this process, a separate tool is mounted either in a 3D printer or a separate gantry system. Wire embedding tools can heat the wire before deposition, thus melting the surrounding plastics during deposition and embedding them within the plastic substrate. A standalone tool or integrated wire heating and delivery tool is needed in addition to a 3D printing machine. In addition to wire co-extrusion and thermal wire embedding tools, ultrasonic welding-based tools are used to deposit wires and carbon fibers within 3D-printed substrates and fabricate hybrid composites. A representative image of the wire embedding process is shown in Figure 5.9 (a) and reported in [10]. The additional tool mounted on the 3D printer for the multifunctional manufacturing of hybrid composites is an inherently slow process. For example, the ultrasonic welding technique shown in Figure 5.9 (b) may not be suitable for fast-rate production in 3D printing. Ultrasonic welding is a slow process requiring precise control [11]. Therefore, adding such tools to the 3D printing machines may require an additional processing chain in the manufacturing process.

Nonetheless, this technique is still emerging, and the applications of such a process for hybrid composites are similar to those in Section 4.2.2. Conductive wire-based composites are used for embedded electronics, sensors, antennae, actuators, and soft robotics applications. Resistive wire-based composites, such as out-of-oven autoclave composite molds, are used in composite tooling.

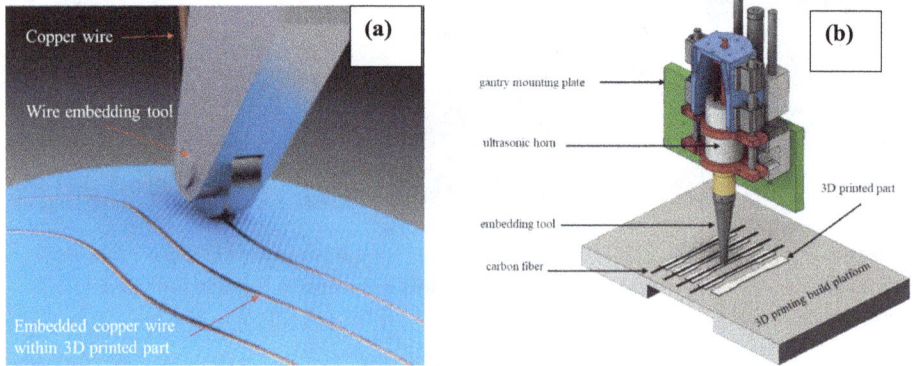

Figure 5.9: (a) Wire-based hybrid composite manufacturing tools used in the FFF system. A conceptual thermal wire embedding tool to deposit copper wires within a 3D-printed substrate [10]. (b) An ultrasonic welding tool fabricated for fiber-reinforced 3D-printed thermoplastic composites [11].

5.4.3 Hybrid composites by multi-material 3D printing

Multi-material 3D printing is a well-known technology for hybrid composites with functionally gradient structures. As described in the previous sections, dual-material 3D printing can be achieved by dual extrusion technology. In addition to dual extrusion, in-process blending in the extruder can also fabricate a functional gradient structure. For example, the ORNL BAAM extruder was modified to deposit neat, fiber-reinforced thermoplastic resin. A customized dual hopper extrusion system was developed at ORNL and adopted in the BAAM machine to enable the manufacturing of blended composite parts [12]. A similar process is being developed for the small-scale FFF 3D printing machine [13] at Pacific Northwest National Laboratory. Multi-material parts fabricated using the dual hopper or feeder in a single mixing nozzle enable the control of the compositional gradient. For example, precise control of the fiber wt% within the thermoplastic resin was possible during the in-situ mixing and deposition. As shown in Figure 5.10 (a), the transition of the neat thermoplastic ABS to the carbon fiber-reinforced ABS was possible in a controllable manner spatially by introducing carbonization in the nitrogen process. Figure 5.10 (b) shows the controlled wt% of the carbon fiber content within the extrudate bead with respect to the location of the print start and switching point.

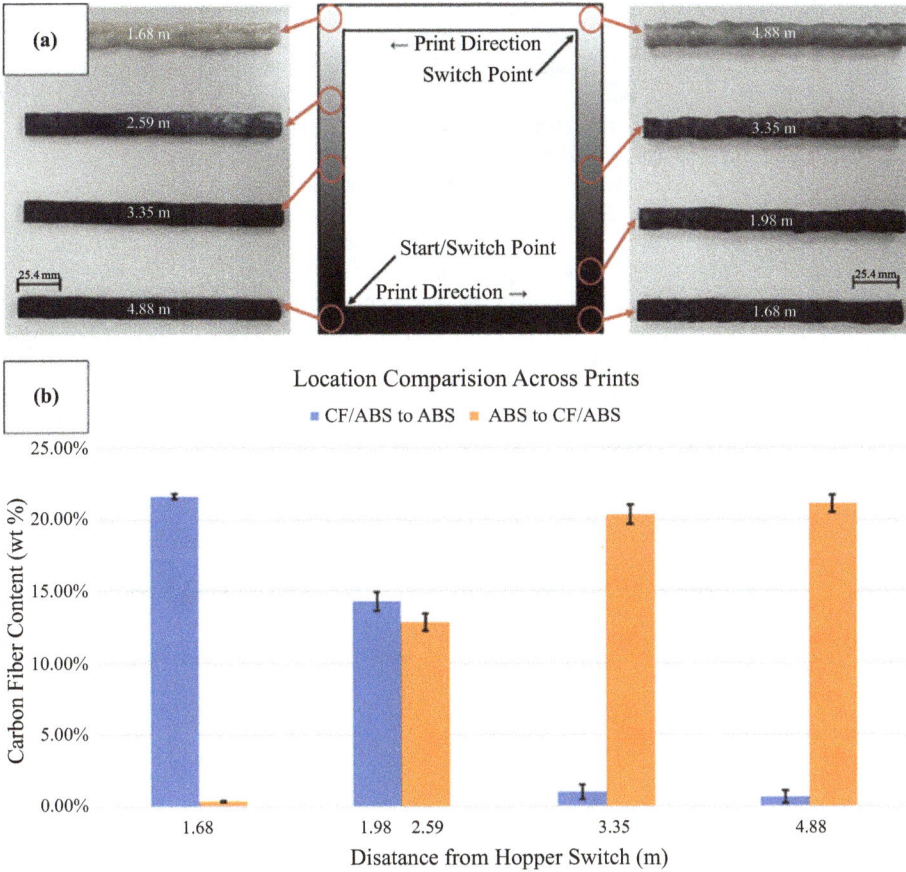

Figure 5.10: (a) Characterization of material transition in BAAM 3D printing with a custom-made dual hopper in a single extruder. Representative sampling locations for the process repeatability portion of the study. The number in the center of each rod measures the distance from the hopper switch. (b) The carbon fiber content at each carbonization in nitrogen testing location for both transition directions [12].

References

[1]   C.E. Duty, V. Kunc, B. Compton, B. Post, D. Erdman, R. Smith, R. Lind, P. Lloyd, L. Love Structure and mechanical behavior of Big Area Additive Manufacturing (BAAM) materials, Rapid Prototyping Journal, 2017, 23(1): 181–189.
[2]   K. Alhassoon, Y. Malallah, F.N. Alsunaydih, F. Alsaleem Three-dimensional printed annular ring aperture-fed antenna for telecommunication andbiomedical applications, Sensors, 2024, 24(3): 949.
[3]   J. Brackett, A. Defilippis, T. Smith, A. Hassen, V. Kunc, C. Duty Evaluating the Mechanical Properties within the Transition Region of Multi-Material Large-Format Extrusion Additive Manufacturing Proceedings of the Solid Freeform Fabrication Symposium, 2022, Austin, Texas.

[4]   R. Johnston, Z. Kazancı Analysis of additively manufactured (3D printed) dual-material auxetic structures under compression, Additive Manufacturing, 2021, 38: 101783.

[5]   M. Rafiee, R.D. Farahani, D. Therriault, Multi-material 3D and 4D printing: A survey, Advanced Science, 2020, 7(12): 1902307. doi: 10.1002/advs.201902307.

[6]   R. Matsuzaki, M. Ueda, M. Namiki et al. Three-dimensional Printing of continuous-fiber composites by in-nozzle impregnation, 2016, Science Report, 6: 23058.

[7]   Y. An, J. Ho Myung, J. Yoon, W.-R. Yu Three-dimensional printing of continuous carbon fiber-reinforced polymer composites via in-situ pin-assisted melt impregnation, Additive Manufacturing, 2022, 55: 102860.

[8]   A. Roschli, M. Borish, B. Post, P. Chesser, J. Heineman, C. Atkins Design for Slicing in Large Format Fused Filament Fabrication, Oak Ridge National Laboratory (ORNL), Oak Ridge, TN (United States), 2019.

[9]   K.M.M. Billah, J. Heineman, P. Mhatre, A. Roschli, B. Post, V. Kumar, S. Kim et al. Large-scale additive manufacturing of self-heating molds, Additive Manufacturing, 2021, 47: 102282.

[10] E. Aguilera, J. Ramos, D. Espalin, F. Cedillos, D. Muse, R. Wicker 3D printing of electro mechanical systems, Proceedings of the Solid Freeform Fabrication Symposium, 2013, Austin, Texas.

[11] K.M.M. Billah, J.L. Coronel Jr, L. Chavez, Y. Lin, D. Espalin Additive manufacturing of multimaterial and multi-functional structures via ultrasonic embedding of continuous carbon fiber, Composites Part C: Open Access, 2021, 5: 100149.

[12] J. Brackett, Y. Zhe Yan, D. Cauthen, V. Kishore, J. Lindahl, T. Smith, Z. Sudbury, H. Ning, V. Kunc, C. Duty Characterizing material transitions in large-scale additive manufacturing, Additive Manufacturing, 2021, 38: 101750.

[13] Z.C. Kennedy, F.C. Josef Printing polymer blends through in situ active mixing during fused filament fabrication, Additive Manufacturing, 2020, 36: 101233.

Steve Bullock, Tony Beard, David Nuttall, Akash Phadatare,
Uday Vaidya

6 Advanced integrated AM filament winding

6.1 Introduction

This chapter on composite fabrication will delve into the filament winding method for continuous fiber composites, a versatile technique when coupled with additively manufactured mandrels. Filament winding (FW) stands out as one of the most cost-efficient methods, with the fiber being directly applied to a preform [1]. No weaving or fabric stitching is required for continuous fiber reinforcement of composite parts. This versatility is evident in the wide range of filament-wound composite parts found in various industrial applications. These parts, which are commonly used in aerospace, defense, automotive, marine, oil and gas, and storage applications, come in a variety of shapes and sizes, demonstrating the adaptability of the filament winding method. Less attention has been given to mandrel design and manufacturing. Depending on the end application, these mandrels are made of metal or thermoplastic liners. Advanced additive manufacturing (AM) has revolutionized composite mandrel production. AM has made an innovation by combining with legacy methods such as filament winding (FW).

6.2 Background

FW consists of a rotating preform called a mandrel mounted on a spindle, on which the fiber is deposited after it traverses a resin path that wets the fiber to provide a matrix that will cure, creating the composite part. Initial parts produced were rota-

Notice: This manuscript has been authored by UT-Battelle, LLC, under contract DE-AC05-00OR22725 with the US Department of Energy (DOE). The US government retains and the publisher, by accepting the article for publication, acknowledges that the US government retains a nonexclusive, paid-up, irrevocable, worldwide license to publish or reproduce the published form of this manuscript, or allow others to do so, for US government purposes. DOE will provide public access to these results of federally sponsored research in accordance with the DOE Public Access Plan (https://www.energy.gov/doe-public-access-plan).

Steve Bullock, Tony Beard, David Nuttall, Manufacturing Science Division, Oak Ridge National Laboratory, Oak Ridge, TN, United States
Akash Phadatare, Tickle College of Engineering, University of Tennessee, Knoxville, Knoxville, TN, United States
Uday Vaidya, Manufacturing Science Division, Oak Ridge National Laboratory, Oak Ridge, TN, United States; Tickle College of Engineering, University of Tennessee, Knoxville, Knoxville, TN, United States; Institute for Advanced Composites Manufacturing Innovation, Knoxville, TN, United States

https://doi.org/10.1515/9783111019543-006

tionally symmetric and usually were limited to two or three axes [2]. It lends itself well to piping and other cylindrical parts such as tanks, casings, and flanges. CAD winding software plays an integral part in the FW process and has allowed this method to create four axes in axisymmetric and non-axisymmetric parts, such as pipe elbows, via winding patterns [3, 4]. These winding patterns are broken into helical, hoop, and polar. Hoop patterns are generally uniform spacing of wetted fiber near 90° from the rotation axis, and helical patterns have changing fiber angles from 20° to 80°. An excellent example of hoop versus polar patterns in composite parts is the manufacture of compressed gas tanks as depicted in Figure 6.1. The ends of the compressed gas tanks, which are near 0°, use polar patterns, and the body of the tank uses helical patterns [5].

The strength of a composite is heavily influenced by the ability to tailor fiber orientation. This is largely determined by the interface between the fiber and the matrix, which plays a crucial role in the matrix's ability to transfer load to the fiber due to interfacial adhesion strength. This strength also affects the resistance to time-dependent deformation under constant load, known as creep. Wound filaments exhibit the highest strength and creep resistance in the direction of the fiber. For instance, winding paths near 90° will have the highest strength but less bending resistance parallel to the winding axis. This is why filament-wound piping often features crisscrossing of fibers to provide the necessary reinforcement for large tanks and pipelines that need to balance tensile forces in all three directions from gas pressure inside.

Figure 6.1: Filament path examples for winding large-diameter tanks; image courtesy of Magnum Venus Products.

One of the characteristics of FW that sets it apart from other methods is the speed at which fiber can be deposited. Depending on the friction effects of the fiber path, fiber can be coated with resin and deposited at the rate of feet per second. Larger carbon fiber tow sizes of 24k or 48k tows are applied at over 100 feet per minute, which translates to up to 1 kg per minute of wetted fiber deposition [6]. The tension on the wetted fiber can be controlled via a tensioner (or creel) system that controls the fiber tension fed through the resin bath. This is useful for controlling the tension and preventing fibers from breaking due to too much strain. Aerospace-grade carbon fiber has strain-to-failure percentages from 1.4% to 2.1% prior to breaking, and stretching fibers during deposition imparts residual stress in the part. The resin bath is a wheel dipped in resin, and dry fiber is converted to wet fiber. This resin wetting is known as wet winding. There are other methods where the fiber is wound dry with the carbon fiber filaments pre-impregnated with resin, called prepregs, which is known as dry winding. The uncoated carbon fibers are pre-impregnated with resin by equipment known as a towpregger. Towpregged carbon fibers are often B-staged, a term that refers to partially cured resins to prevent the liquid from running out of the fiber, with enough residual cure to form a strong composite after winding and curing at recommended curing profiles.

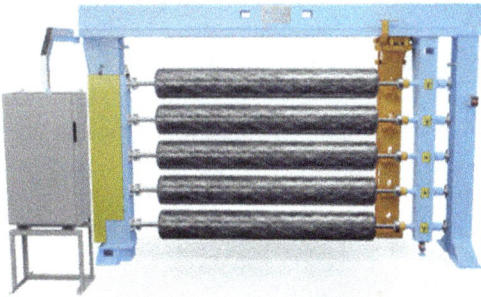

Figure 6.2: Multiaxial filament winder. Image courtesy of Magnum Venus Products.

Winding patterns can be calculated and considered, including the bandwidth of the fiber, i.e., the spread width each fiber filament makes when pressed against the mandrel. The fiber's bandwidth and yield are used to determine the coverage percentage. The winding calculation also factors in the friction limit, corresponding to the limit and speed at which slipping occurs over the mandrel. The slipping can depend on the angle of the mandrel, which is very important when winding tanks. Finally, structural considerations and ultimate stresses will determine the thickness of the filament-wound part.

Filament winding is easily scaled to support the rapid production of composite parts. Multiple axes of the FW system, as shown in Figure 6.2, demonstrate the scalability and flexibility of FW for producing composite tanks, small piping, and pipelines. Spacing the mandrels far apart can allow winding tanks or pipes up to 16 feet in diameter and 40 feet long. Potential applications include liquefied natural gas (LNG) tanks. Like all other FW applications, the wetted carbon filament is wound over a mandrel. The mandrel must be removed before the final part can be used. A mandrel can be a metal pipe and rubber bladder that is inflated, wound with fiber, and deflated to facilitate mandrel removal.

AM has been combined with other legacy composite processes, such as AM with compression molding [7]. Combining manufacturing techniques to create hybrid approaches can increase part production rates, so a natural progression is to find other techniques that are compatible with the hybrid approach. AM and filament winding are discussed as such a combination of legacy – FW with AM. Hybrid approaches discussed in this chapter will incorporate mandrels that are additively manufactured with thermoplastics or thermoset materials.

6.3 Hybrid filament winding with AM mandrels

Mandrels for FW systems can be manufactured via additive manufacturing , offering faster production and less machining than their metallic counterparts. This hybrid manufacturing technique, when coupled with FW, provides the best of both worlds in terms of logistical and processing timeframes for rapid production rates. The types of additive manufacturing for mandrels, which use polymer deposition, are divided into two categories: thermoplastic and thermoset. The thermoplastic mandrels can employ a variety of equipment, from fused deposition modeling (FDM) to thermoplastic extruder systems, such as the big area additive manufacturing (BAAM) system that uses thermoplastic pellet feedstock to print mandrel shapes.

Figure 6.3: Additively manufactured mandrel; Image courtesy of ORNL.

Figure 6.3 shows an FDM-printed mandrel on a spindle mounted in the chucks of a filament winder, ready to be wound. Printing the part, mounting the mandrel, and winding take less than a day. Thermoplastic mandrels use extruders to melt thermoplastic pellets (see Figure 6.4) into the desired mandrel shape. The tool path for the extruder, used to create the shape, is generated by CAD files, known as STL files. These files, the predominant type when generating an AM part, provide the 3D point location for the extruder to follow. Tool pathing is determined by various software tools that convert G-code to a path that controls machine movement for depositing thermoplastic and thermoset materials.

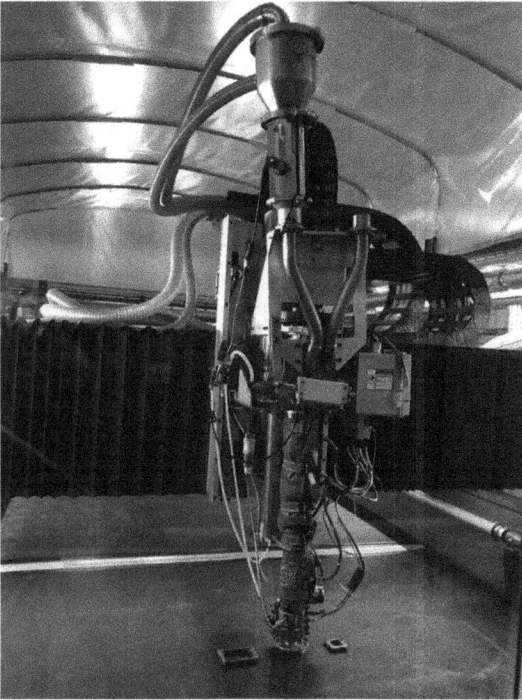

Figure 6.4: Thermoplastic extruder for BAAM AM mandrel; image courtesy of ORNL.

Additively manufactured thermoset mandrels can increase production rates using accelerated cures for epoxies in standard filament winding. Accelerated cures incorporating UV curing can form mandrels from polyesters and epoxies that cure in minutes. This rapid mandrel deposition using UV-cured epoxies and polyesters can create mandrels from new designs that take days instead of months from design inception to AM printing and filament winding. Thermoset printing systems consist of one-part (called 1 K) and two-part (2 K) resin systems. Typical resins used in 2 K systems include urethanes, epoxies, and polyesters. Polyesters are often used in the marine and automotive industries due to their low price and high performance. Another benefit of AM mandrels is weight. Metal mandrels that are up to 4 feet long and 2 feet in diameter would be heavy to lift, and plastic mandrels can be filled with lightweight reinforcing fillers such as glass microspheres to improve toughness and reduce the weight of AM mandrels compared to their metallic counterparts. Fillers such as Teflon flake can be added to AM mandrels, primarily to thermoset AM mandrels, to aid in removing the mandrels from the filament-wound part after curing.

6.4 Applications of AM-FW composites

As discussed earlier, another area where AM could be helpful is the rapid manufacturing of custom mandrels. AM's capability to produce complex shapes helps reduce the total weight of the mandrel and part manufacturing time using FW. The researchers at ORNL are trying to modify the FW mandrel using AM due to its inherent characteristics of custom-built parts. The idea is that an AM-printed geodesic cylindrical mandrel would save the mandrel material and make it lightweight. The printing material is a water-dissolvable thermoplastic material, giving the freedom to dissolve the complete material after the FW step, further reducing the weight of the part. This approach could be beneficial in many ways, including: energy-efficient FW as the motor has to rotate a lightweight mandrel instead of a completely printed one; faster FW part production as low mandrel material results in lesser dissolution time after the FW operation, etc. In the current demand for sustainable manufacturing techniques, the integration of AM in FW has resulted in one such hybrid production method. Figure 6.5 represents the geodesic AM-printed mandrel alongside the traditional entirely printed mandrel after hoop winding. There is vast potential for using different topologically optimized designs for the mandrel to produce FW parts with tailored local properties.

Another compelling example of hybrid AM/FW technology can be seen in the manufacturing of nonsymmetrical parts used in the construction field, as shown in Figure 6.6 [8]. Here, the enclosure caps are made by AM with protrusions for FW [9]. A specialized FW, robotic coreless filament winding (RCFW), is used for this operation [10]. In this case, it is critical to define the toolpath for the robotic arm due to the nonsymmetrical nature of the parts to be produced using RCFW. The biggest benefit gained by this technique is the manufacturability of complex nonsymmetrical shapes, which is impossible with usual methods. This underscores the need for further research to synergistically combine AM with FW to realize many similar applications described earlier, offering hope for the future of this technology.

In another application, a practical challenge for the AM-FW technology is presented in the construction of a medium-sized filament-wound drone motor. The inertial forces of the spinning portions of the motor must be maintained to avoid failure of the rotor components. The plausibility of 3D printing a mandrel was validated with the fabrica-

Figure 6.5: Hoop winding on a geodesic cylindrical mandrel (left) and a conventional cylindrical mandrel (right); image courtesy of ORNL.

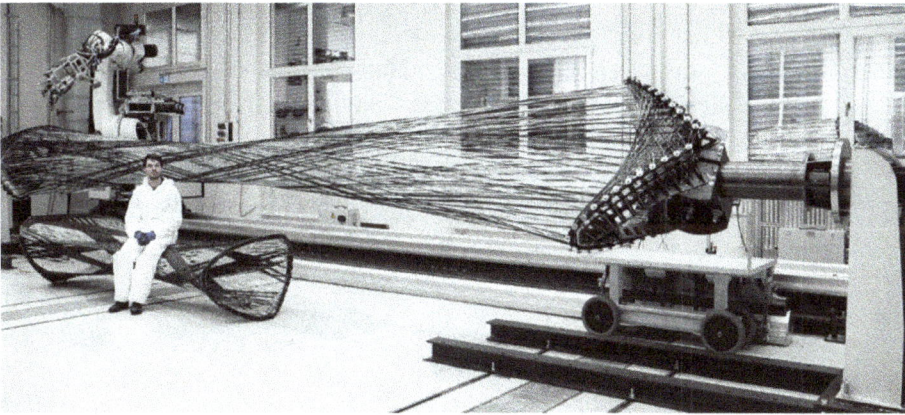

Figure 6.6: Demonstration of complex, nonsymmetrical composite parts manufactured using RCFW [8].

tion of a mandrel foundation from pelletized ABS/CF. After short-term cooling and curing, the mandrel was hoop-wound with a small-scale X-winder filament winder.

Taking that knowledge forward, a functioning drone motor was disassembled, and the rotor was reverse-engineered for dimensional purposes. 3D magnets, Fe/Nd/B in Nylon 12, were printed and poled N-S in alternating fashion in a 5 Tesla magnetic field, which was heated to 167° C to allow softening of the Nylon 12 for the alignment of magnet powder in the field to allow the motor operation. Poled magnets were epox-

(a)   Poled Magnet Design   (b) Magnet Holder   (c) Magnet Holder (d) Heated Magnets (d) Poled/Bonded Magnets

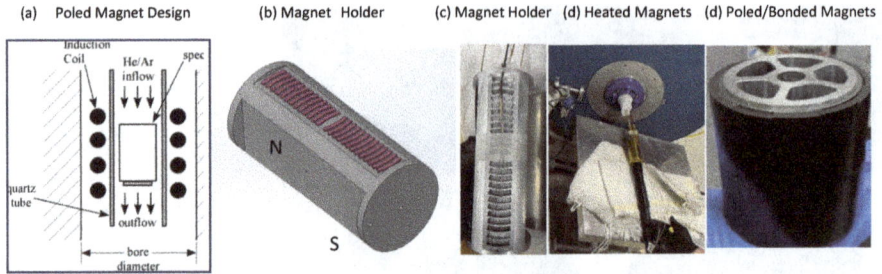

Figure 6.7: AM magnet printing and poling process. Image courtesy of ORNL.

ied to the mandrel shape, followed by hoop winding with epoxy-impregnated continuous carbon fiber, and oven-cured to set the epoxy (see Figure 6.7). The outer surface of the rotor was machined to the appropriate tolerance, spin-balanced for minimum vibration, and finally assembled into the original motor stator and casing (see Figure 6.8). Mechanically, that sufficed to validate the process of creating AM-FW motor components. This successful validation of the AM-FW motor components should instill confidence in the capabilities of this technology. However, that did not satisfy the reproduction of a functional motor until it was taken to the test stage and performance validated.

Partnering with a Crossville, TN company the authors functionally verified the motor performance through a series of tests, culminating in a multistep increase topping out at 6,000 RPM. The motor was inspected, and no mechanical performance issues were found.

Figure 6.8: AM with UV cure resin capability (right) installed with FW (X-winder); image courtesy of ORNL.

6.5 Future challenges with AM-FW solutions

Filament winding combined with AM has focused on traditional winding aspects onto mandrels printed via AM. The fibers are added as layers via different filament orientations, and the Z-axis strength primarily relies on resin adhesion to the deposited layer. When exposed to out-of-plane loading, filament-wound parts often suffer from lower interlaminar shear strength (ILSS) and fracture. The advent of computer-controlled robotic manufacturing allows Z-axis reinforcement to improve delamination resistance. While Z-axis reinforcement afforded by robotic manufacturing improves the ILSS, there is a penalty for improving ILSS. Tensile strength is reduced due to stitching breaking fiber in the plane [11]. The degree of stitching must be balanced. Greater spacing results in less reduction in tensile compressive strength and improvements in the ILSS and fracture toughness. One of the drawbacks to stitching in manually stitched filament-wound composites is the high labor costs. The motivation for incorporating robotic hybrid manufacturing is to bring down costs. The following section will discuss several aspects of Z-axis reinforcement.

An area of research related to FW and, inherently, AM-FW involves the modification of the Z-axis (through thickness) fiber architecture. This area of research aims to improve the composite's ILSS while minimizing the degradation of other mechanical properties in the axial direction. The methods employed to improve ILSS vary but are generally divided into native fiber reorientation and foreign material reinforcement. See Table 6.1 for a delineation of these methods.

Table 6.1: Through thickness fiber modification methods [12].

Native fiber reorientation method	Mechanism	Non-native material introduction	Mechanism
Needle-punching [13]	Barbed needles are rapidly cycled through the composite layup, damaging some of the fibers while simultaneously interlocking each ply's fibers together.	Stitching	The composite layup is stitched (sewn) together with a foreign fiber using a bobbin and needle thread.
		Tufting	Like stitching, except only one sewing thread is advanced through the layup thickness and is retained by friction between the plies of the laminate.
		Z-pinning	Hardened metallic or composite pins are advanced through the layup thickness and co-bonded to the layup.

Details of the ongoing research in through-thickness composite reinforcement are beyond the scope of this chapter but are mentioned to communicate the synergistic technology efforts underway in the composite manufacturing industry. One area of composite manufacturing that incorporates through-thickness reinforcement is carbon preforms used in carbon-carbon (CC) and carbon-ceramic (CMC) brake pads and rotors. CC and CMC brake materials are valued for their ability to withstand extremely high temperatures without wear or performance loss. They are used to manufacture aircraft brakes and in advanced racing applications. The pads and rotors are typically fabricated using carbon felt and subsequently needled together before converting to high-temperature composites [14]. The needling action causes fiber entanglement between each ply of felt, mechanically binding each ply together. This reinforces the chemical bonding between each ply and the composite matrix during high-temperature processing.

Each of the through-thickness reinforcement techniques mentioned above presents a unique set of challenges. For instance, within AM-FW, current research efforts are focused on the optimization of AM preforms. This optimization is necessary to allow the needles to penetrate through the layup thickness without damaging the underlying mandrel or the needles. This ongoing effort underscores the complexity and dedication involved in the research and development of through-thickness reinforcement techniques.

6.6 Conclusion

Hybrid manufacturing is a continuing innovation that couples legacy composite techniques with additive manufacturing. The goal is greater efficiency, hence lower costs, by combining complementary processes, such as AM-FW. AM-FW allows for rapid design iteration made feasible by AM, while using the high throughput of FW. A unique aspect is the relative ease of making FW preforms as opposed to legacy methods such as machining an aluminum mandrel. The fast design iteration coupled with rapid deposition with a thermoset printer allows the hybrid AM-FW process to occur faster than separate techniques. Reducing touch labor costs is an additional benefit; without the need to transfer parts from one location or facility to another, this co-located aspect saves valuable time from a manufacturing flow viewpoint. Unique applications such as motor and rotor casings for electric motors are one such example. The future of AM-FW would lend itself to other cutting-edge technology, such as insulated compressed gas cylinders for liquid cryogen storage. The AM portion can print foam insulation followed by traditional FW of the printed insulation. The potential applications lend themselves to the energy and aerospace sectors.

Acknowledgements: The authors acknowledge the support from the US Department of Energy (DOE), Office of Energy Efficiency and Renewable Energy, and Advanced Materials and Manufacturing Office. This manuscript has been authored by UT-Battelle, LLC, under contract DE-AC05–00OR22725 with the US Department of Energy (DOE). The US government retains and the publisher, by accepting the article for publication, acknowledges that the US government retains a nonexclusive, paid-up, irrevocable, worldwide license to publish or reproduce the published form of this manuscript, or allow others to do so, for US government purposes. DOE will provide public access to these results of federally sponsored research in accordance with the DOE Public Access Plan (http://energy.gov/downloads/doe-public-access-plan).

References

[1] A.M. Shibley Filament Winding,in: Handbook of Composites, Springer US, Boston, MA, 1982, pp. 449–478. doi: 10.1007/978-1-4615-7139-1_16.

[2] M. Quanjin, M.R.M. Rejab, M.S. Idris, B. Bachtiar, J.P. Siregar Harith MN. Design and optimize of 3-axis filament winding machine, IOP Conference Series: Materials Science and Engineering, 2017, 257: 012039. doi: 10.1088/1757-899X/257/1/012039.

[3] A. Andrianov, E.K. Tomita, C.A.G. Veras Telles B. A low-cost filament winding technology for university laboratories and startups, Polymers (Basel), 2022, 14(5): 1066. doi: 10.3390/polym14051066.

[4] T. Sofi, S. Neunkirchen, R. Schledjewski Path calculation, technology and opportunities in dry fiber winding: A review, Advanced Manufacturing: Polymer & Composites Science, 2018, 4(3): 57–72. doi: 10.1080/20550340.2018.1500099.

[5] S. Peters Composite Filament Winding, ASM International, Materials, Park, OH 44073, 2011.

[6] G. Gardiner Filament winding, reinvented, Compos World, 2018, 24–27.

[7] V. Kumar, S.P. Alwekar, V. Kunc et al High-performance molded composites using additively manufactured preforms with controlled fiber and pore morphology, Additive Manufacturing, 2021, 37: 101733. doi: 10.1016/j.addma.2020.101733.

[8] S. Bodea, P. Mindermann, G.T. Gresser, A. Menges Additive manufacturing of large coreless filament wound composite elements for building construction, 3D Print Additive Manufacturing, 2022, 9(3): 145–160. doi: 10.1089/3dp.2020.0346.

[9] P. Mindermann, G.T. Gresser Adaptive winding pin and hooking capacity model for coreless filament winding, Journal of Reinforced Plastics and Composites, 2023, 42(1–2): 26–38. doi: 10.1177/07316844221094777.

[10] P. Mindermann, S. Bodea, A. Menges, G.T. Gresser Development of an impregnation end-effector with fiber tension monitoring for robotic coreless filament winding, Processes, 2021, 9(5): 806. doi: 10.3390/pr9050806.

[11] R. Velmurugan, S. Solaimurugan Improvements in Mode I interlaminar fracture toughness and in-plane mechanical properties of stitched glass/polyester composites, Composites Science and Technology, 2007, 67(1): 61–69. doi: 10.1016/j.compscitech.2006.03.032.

[12] I. Gnaba, X. Legrand, P. Wang, D. Soulat Through-the-thickness reinforcement for composite structures: A review, Journal of Industrial Textiles, 2019, 49(1): 71–96. doi: 10.1177/1528083718772299.

[13] B.D. Lawrence, T.A. Bogetti, R.P. Emerson Processing and characterization of needled carbon composites, CAMX 2015 – the Composites and Advanced Materials Expo, 2015, 2679–2695.

[14] E.L. Renaud Duval Method of manufacturing carbon-carbon composite brake disks, 2001, 1(12).

Ahmed Arabi Hassen, Segun Isaac Talabi, Vipin Kumar

7 Innovations in high-rate composite manufacturing: integrating additive manufacturing with compression molding process

Abstract: Advanced composites play a pivotal role in modern engineering, offering exceptional strength-to-weight ratios and tailored properties, essential for various industries. High-rate composite manufacturing techniques have rapid production capabilities, which are essential for meeting the demands of industries requiring cost-saving, efficiency, and quick turnaround times. This chapter explores the Additive Manufacturing-Compression Molding (AM-CM) system developed by Oak Ridge National Laboratory (ORNL) for advanced composites manufacturing. The AM-CM system integrates additive manufacturing with compression molding, facilitating the production of polymer composite parts with superior mechanical properties and meticulously controlled microstructures. This innovative system not only ensures precise material deposition but also operates as a fast composite manufacturing process, enhancing productivity and performance, which are needed attributes across industrial applications. Through comprehensive mechanical testing and microstructural analysis, AM-CM promotes remarkable fiber alignment and reduced porosity in composite parts compared to alternative thermoplastic high-rate composite manufacturing methods. Furthermore, AM-CM enables overmolding reinforcement using continuous carbon fiber and supports selective reinforcement through customizable toolpaths. It also facilitates the production of hybrid materials to achieve tailored mechanical properties. Future advancements in AM-CM technology aim to enhance process efficiency, broaden material versatility, and improve part performance. This involves exploring novel materials, advancing process monitoring, implementing automation technologies, and integrating artificial intelligence (AI) and machine learning (ML) for predictive modeling and real-time optimization in composite manufacturing. These developments will establish the AM-CM system as a transformative technology in composite manufacturing, driving innovation across industries.

Notice: This manuscript has been authored by UT-Battelle, LLC, under contract DE-AC05-00OR22725 with the US Department of Energy (DOE). The US government retains and the publisher, by accepting the article for publication, acknowledges that the US government retains a nonexclusive, paid-up, irrevocable, worldwide license to publish or reproduce the published form of this manuscript, or allow others to do so, for US government purposes. DOE will provide public access to these results of federally sponsored research in accordance with the DOE Public Access Plan (https://www.energy.gov/doe-public-access-plan).

Ahmed Arabi Hassen, Segun Isaac Talabi, Vipin Kumar, Manufacturing Science Division, 2350 Cherahala Blvd, Oak Ridge National Laboratory, Oak Ridge, Knoxville, TN 37932, USA

https://doi.org/10.1515/9783111019543-007

Keywords: advanced composites, fiber alignment, high-rate manufacturing, reduced porosity, mechanical properties, AM-CM system

7.1 Introduction

Advanced composites are critical in modern engineering. These materials have good strength-to-weight ratios and tailored properties. These materials are used in many applications, from airplane parts to lightweight car parts, and allow for new designs in many industries [1]. Advanced composites are versatile and adaptable, making them essential for progress in many fields. Despite the remarkable strides that have been made in advanced composite manufacturing, significant challenges persist. These challenges include cost, scalability limitations, and the complexity of optimizing the manufacturing process [2, 3]. However, within these challenges lie opportunities for innovation and collaboration.

High-rate composite manufacturing techniques such as injection molding, compression molding, and thermoforming are suitable for applications in the automotive, commodity, defense, aerospace, and consumer goods industries [4–7]. For example, in the automotive sector, they can be used to produce components such as seat backs and battery trays [8, 9]. High-rate composite manufacturing offers significant advantages that address modern industrial demands. These techniques enable rapid production of complex shapes and structures without compromising quality [10]. This efficiency reduces manufacturing lead times and overall costs, making high-rate methods economically advantageous. Additionally, these techniques often involve automated processes that ensure consistency and repeatability, minimizing human error and increasing product reliability [11]. Moreover, high-rate manufacturing can support scalable production volumes, making it ideal for industries that require large quantities of composite components. However, the substantial initial investment needed for specialized equipment and facilities is a major drawback of these techniques [12, 13].

Understanding the influence of polymer selection is pivotal in addressing these challenges. Two primary types of polymers utilized in composites are thermosets and thermoplastics. Thermoset polymers feature cross-linked molecules, forming a network that does not soften when heated, but decomposes. Thermosets can be utilized in high-rate manufacturing through Sheet Molding Compounds (SMC) or Bulk Molding Compounds (BMC). SMC, composed of thermoset resin, fillers, and chopped glass fibers, is utilized in compression molding to produce large, complex parts with high strength and rigidity, such as automotive body panels and industrial equipment housings [14–17]. BMC, a blend of thermoset resin, short glass fibers, fillers, and additives, is used in injection and compression molding to create intricate, precise components like electrical housings and appliance parts. Both materials offer excellent mechanical properties, dimensional stability, and resistance to corrosion and heat, making them ideal for various demanding applications.

On the other hand, thermoplastic polymers consist of linear chain molecules that enable them to soften and be reprocessed multiple times. Thermoplastic composites offer several advantages over thermoset materials, including recyclability, shorter processing times, absence of exothermic reactions or toxic emissions, and superior damping capacity [18, 19]. These materials provide a unique combination of high strength, stiffness, impact resistance, and design flexibility, making them desirable for various applications. Thermoplastics can be processed through extrusion, compression molding, or injection molding, and can also be used in sheet forms such as organosheets, films, membranes, pipes, and thermoplastic prepregs [9, 20–23]. Their recyclability makes thermoplastics environmentally friendly and cost-effective, with shorter processing times [24]. Their superior damping capacity and ability to be molded into complex shapes further enhance their desirability in high-rate manufacturing applications [25, 26].

The properties of a composite are influenced not only by the type of polymer resin but also by the form and arrangement of the reinforcing fibers. In polymeric composites, reinforcements can be categorized into two main groups: continuous reinforcement and discontinuous reinforcement. This distinction is based on the fiber length or aspect ratio (length-to-diameter ratio). In this chapter, we will focus on discontinuous reinforcement. Discontinuous fiber-reinforced composites have aspect ratios that range from 4 to 2000. The main aspects that influence the composites' properties are a) fiber length, b) fiber orientation, c) fiber–matrix interface, d) porosity, and e) fiber volume fraction.

a. *Fiber Length:* These composites are classified into Short Fiber Thermoplastics (SFTs) and Long Fiber Thermoplastics (LFTs) based on the fiber aspect ratio. SFTs have aspect ratios below 100, while LFTs range from 100 to 2000. SFTs are used to improve the mechanical properties of the unfilled polymer, even though their fiber length is less than the critical fiber length. On the other hand, LFTs have fiber lengths equal to or greater than the critical fiber length (see Figure 7.1). With high fiber alignment, LFTs' mechanical properties can reach those of continuous fiber composites [9]. This makes LFTs particularly desirable because of their ability to be processed quickly into complex shapes.

b. *Fiber Orientation:* Fiber orientation is crucial in determining the thermal, mechanical, and thermomechanical properties of composite materials. Equations such as the rule of mixtures, the Halpin-Tsai model, micromechanical models, and finite element analysis can be used to calculate the stiffness of the composites [27]. However, controlling fiber orientation can be challenging. Any change in fiber orientation results in altered mechanical properties, making it essential to understand and control fiber orientation to accurately predict the performance of fabricated composite parts. Figure 7.2 illustrates the effect of fiber orientation on stiffness. The fiber orientation is predefined in continuous fiber composites, allowing for precise prediction of the final mechanical properties based on this predefined information. In contrast, discontinuous fibers processed through injec-

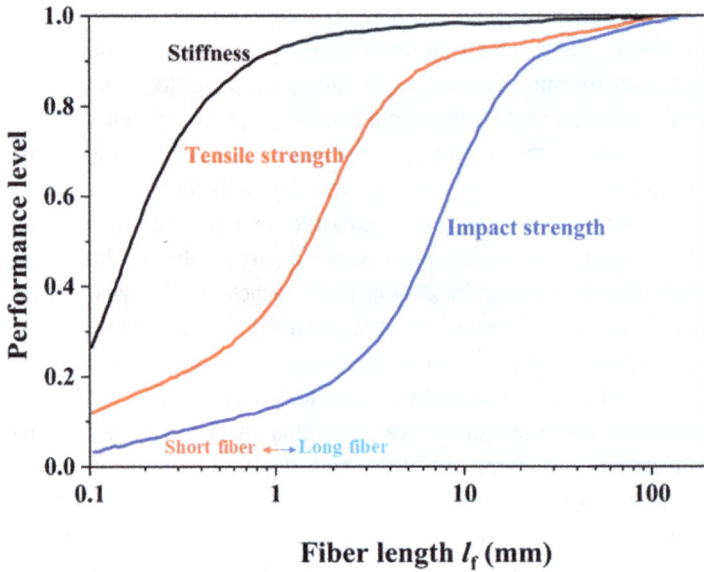

Figure 7.1: Effects of fiber length on stiffness, tensile strength, and impact strength.

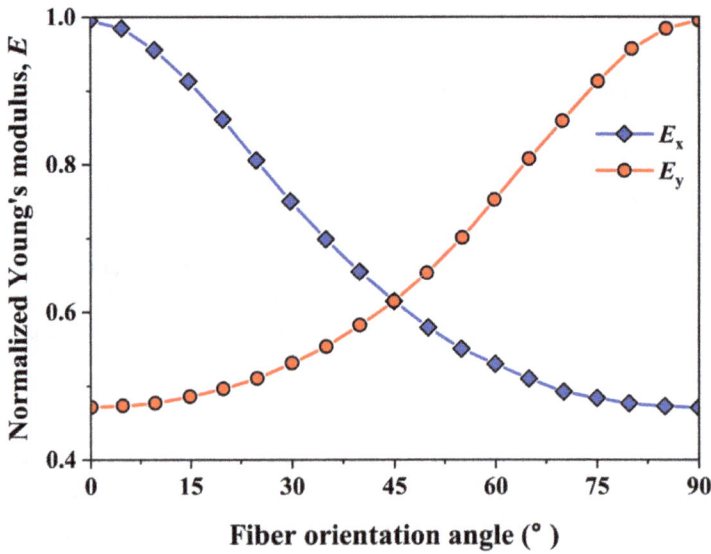

Figure 7.2: The relationship between Young's modulus of composites and the orientation angle of fibers (0° indicating alignment along the fiber direction and 90° indicating alignment perpendicular to the fiber direction).

tion or compression molding do not have a predefined orientation, with their preferred alignment depending on the melt or charge flow direction. This results in an uncontrolled and highly complex fiber orientation, with fibers potentially achieving high alignment in some areas and random orientation in others.

c. *Fiber–Matrix Interface:* The interface is the region where forces are transferred between the reinforcement and the matrix. An effective interface allows for efficient force transfer, which enhances the overall properties of the composite. A well-formed interface contributes to the composite's strength, stiffness, and impact resistance [28, 29]. When the interface is strong, the fibers can effectively carry and distribute the load across the matrix, leading to a more robust and durable material. Adequate fiber sizing is crucial for achieving an effective fiber–matrix interface. Fiber sizing involves coating the fibers with a material that promotes adhesion to the matrix. This coating ensures the fibers are adequately bonded to the matrix, preventing issues such as fiber pullout or debonding under stress. As a result, the composite exhibits improved mechanical properties. Studies have shown that optimized fiber sizing can significantly enhance these properties, making it a critical factor in designing and manufacturing high-performance polymer composites [30].

d. *Fiber Volume Fraction:* There is a critical value of the fiber volume fraction below which the composite strength is primarily governed by the matrix stress at the point of composite failure. This critical value defines the minimum fiber volume fraction necessary for the reinforcement to effectively enhance the composite's mechanical properties. The mechanical properties of the composite are directly proportional to the fiber volume fraction up to this critical point. However, exceeding a maximum limit for the fiber volume fraction can lead to insufficient wetting of the fibers by the matrix, resulting in reduced mechanical properties. The relationship can be described by the matrix stress at composite failure, a shape parameter, and the fiber's ultimate stress. Proper balance and optimization of fiber volume fraction are essential for maximizing the reinforcement efficiency and overall performance of the composite material.

This chapter discusses the pioneering efforts in developing an integrated manufacturing system that combines controlled material deposition to tailor properties with the advantages of high-rate manufacturing processes for improved performance. This innovative approach aims to enhance the mechanical properties and efficiency of composite materials by leveraging advanced manufacturing techniques. The chapter covers the background and rationale for the innovation, concept proofing, understanding of material microstructure, and system development. It also provides an outlook on system development and potential application spaces.

7.2 Integrated Additive Manufacturing– Compression Molding (AM-CM) Process

7.2.1 Background

Extrusion deposition additive manufacturing

Additive Manufacturing has emerged as a promising method for composite fabrication. This technique revolutionizes the production of complex geometries with minimal material waste, offering unparalleled design freedom and customization capabilities [31]. Utilizing composite filaments or pellets, additive manufacturing methods such as Fused Filament Fabrication (FFF) and extrusion-based deposition [32] enable layer-by-layer deposition of materials to produce complex geometries, offering customization and reduced material waste [33–36]. This is particularly evident in processes like the Big Area Additive Manufacturing (BAAM) system pioneered by ORNL, which excels in creating composite structures with high anisotropy [37]. By leveraging its extrusion-based approach, BAAM achieves high fiber alignment along the deposition direction, which is crucial for enhancing structural integrity and performance in specific applications. The system's capability for multimaterial deposition further enhances its versatility, allowing for precise material placement and design optimization, thereby reducing weight and cost in large-scale components. Nowadays, large-scale AM can produce parts up to 29 m (96 ft) long by 10 m (32 ft) wide by 5.5 m (18 ft) high, with deposition rates reaching up to 227 kg/h (500 lb/h) [38]. As advancements continue, particularly with the incorporation of reactive thermosetting polymers and continuous fiber reinforcements, large-scale AM systems are poised to redefine the landscape of industrial manufacturing [39, 40].

Effect of manufacturing processes on the architecture of discontinuous fiber-reinforced composites

Several factors impact the properties of discontinuous composites, with fiber length distribution (FLD) and fiber orientation distribution (FOD) being foremost among them. In SMC, the fiber length is controlled by adjusting the cutter length size, which affects the uniformity and strength of the final product. In thermoplastic extrusion processes, fiber length is managed through the design of the extruder screw. The screw's profile and configuration, including the compression and mixing zones, influence the shear rate and residence time of the polymer melt and fiber mixture. Lower shear rates and longer residence times typically retain longer fiber lengths in LFT composites. However, this approach may encounter challenges such as mixing issues and the potential for fiber clumping or logging (bridging), which can affect the composite's integrity and performance.

FOD significantly influences the mechanical response of composites. When discontinuous fibers are aligned in a preferred direction, the elastic modulus in the longitudinal direction is lower than that of continuous fibers, and it is affected by the fiber aspect ratio. This alignment provides substantial reinforcement along the fiber direction, enhancing the composite's strength and stiffness. When discontinuous fibers are randomly oriented, the composite behaves isotropically, exhibiting uniform properties in all directions. This isotropic behavior results in reduced tensile modulus compared to composites with aligned fibers. Random orientation diminishes the reinforcement effectiveness, leading to lower overall strength and stiffness. Proper control over FOD can significantly improve performance. In applications requiring specific directional strength, ensuring fibers are aligned in the desired direction is essential.

In high-rate manufacturing processes such as injection molding and compression molding, discontinuous fibers do not have a predefined orientation. Their alignment is influenced by the melt or charge flow direction, resulting in a complex orientation where fibers may be highly aligned in some areas and randomly oriented in others. In compression molding, altering the charge placement or changing the mold size can help control fiber orientation in the final part [41]. Similarly, adjusting the runners and gate locations in injection molding can influence melt flow to achieve the desired fiber orientation. In both compression molding and injection molding, fiber orientation through the thickness of the composite material varies due to flow dynamics. The surface layers typically have fibers aligned parallel to the mold surfaces because of high shear rates.

In contrast, the core layer exhibits more random orientation or fibers aligned perpendicular to the flow direction due to lower shear forces (see Figure 7.3). This creates a layered structure, with aligned fibers at the surfaces and more varied orientation in the core, impacting the overall mechanical properties of the composite. These traditional high-rate molding processes often lack precise control over fiber orientation and microstructure, limiting the optimization of mechanical properties. Enhancing control over FOD in these manufacturing methods is crucial for achieving superior performance in thermoplastic composites.

In extrusion-based additive manufacturing processes, the manufactured parts demonstrate superior properties in the direction of the deposition [36]. This is primarily due to the high alignment of fibers induced along the deposition path by the flow of material through the deposition nozzle [43–45]. This alignment significantly enhances the strength and stiffness of the part in the direction of deposition, ensuring optimal mechanical performance where it is most critical. The digital nature of additive manufacturing processes allows for precise control over material deposition.

Extrusion deposition additive manufacturing has some drawbacks, such as high porosity, weak properties through the interface (layer-to-layer adhesion), and being relatively slow compared to conventional high-rate manufacturing processes. Porosity can weaken structural integrity and compromise the performance of the composite, especially in applications where strength and durability are crucial. Furthermore,

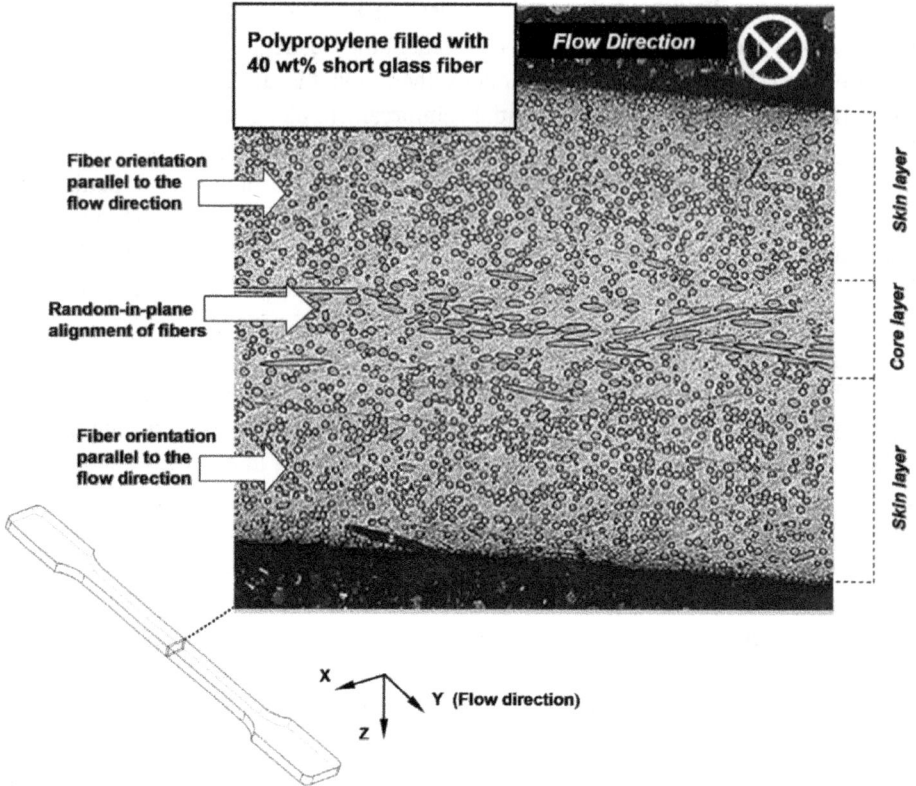

Figure 7.3: Optical micrographs of injection-molded short-fiber polypropylene reveal the fiber orientation pattern in the *z-x* cross-sectional plane [42].

achieving high-quality surface finishes remains a persistent obstacle. The layer-by-layer deposition process often results in visible layer lines and rough surfaces, which may require post-processing treatments to meet desired standards. On the other hand, traditional high-rate manufacturing techniques offer the advantage of producing composite parts at high volumes and speeds, with relatively low porosity and excellent surface finish.

7.2.2 Concept development

ORNL pioneered the integration of AM with CM, introducing the additive manufacturing-compression molding (AM-CM) system. By depositing pre-aligned fibers using AM and subsequently compressing them, this process achieves highly consolidated parts with controlled fiber orientation and minimal porosity [36]. This innovative approach leverages the advantages of controlled fiber alignment in additive manufacturing with the high-volume capabilities, low porosity, and excellent surface finish of traditional compression molding [28].

The conception of this method began with utilizing an existing large-scale additive manufacturing system to print preforms, which were then transferred and heated in a mold using a fast-acting compression molding press, as shown schematically in Figure 7.4.

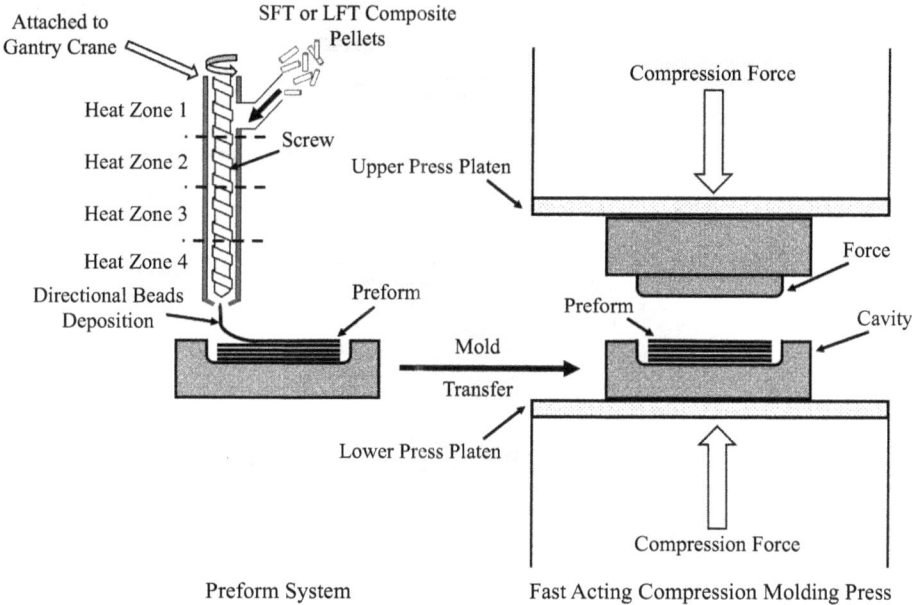

Figure 7.4: Schematic of the developed concept for a tailored preform system utilizing Schematic AM-CM (Courtesy of Oak Ridge National Laboratory).

Figure 7.5 shows the process of depositing PA6/GF 40% by weight to form two different preforms: a) Unidirectional preform (i.e., two layers of deposition in the 0° direction), and b) Cross preform (i.e., two layers, one in the deposition of the 0° direction and another in the deposition of the 90° direction). The extruder starts to travel in the x-direction of the mold, laying down materials and creating the first layer of the preform (i.e., 0° layup). After depositing the first layer of the preform, the extruder moves up in the z-direction and starts to travel in the y-direction, creating another layer of deposited material (i.e., 90° layup), to provide a final part with orthotropic behavior. The preforms were transferred to a 100-ton compression molding machine, as shown in Figure 7.5, which was used to press the preform material.

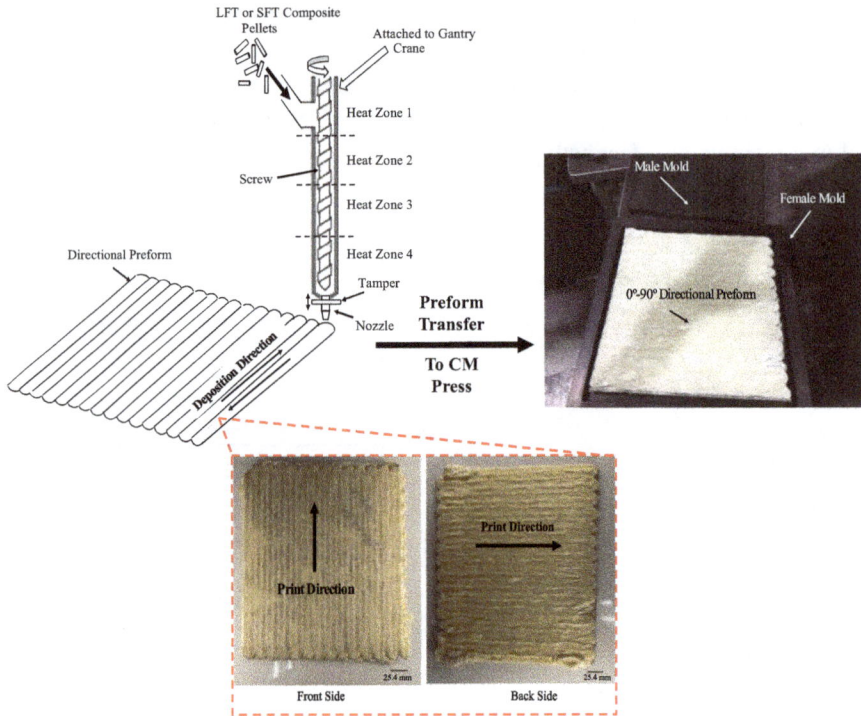

Figure 7.5: Illustration of the concept of the AM-CM process: from material preform via AM to the final parts transferred to the compression molding process (Courtesy: Oak Ridge National Laboratory, USA).

The resulting consolidated panels, shown in Figure 7.6, demonstrate the effectiveness of this integrated manufacturing process in achieving well-consolidated parts with controlled fiber orientation and minimal porosity.

Figure 7.7 shows the microstructure of the cross section of the consolidated panels after compression molding. In the 0°/0° sample, all fibers are aligned in the direction of the deposition, evident from their circular cross sections across the entire thickness. In the 0°/90° sample, one layer displays fibers with circular cross sections, while the 90° layer shows fibers with more elongated ellipsoids and rectangular shapes. This indicates effective control over fiber deposition and orientation through the part thickness. Figure 7.8 presents the mechanical performance of both samples compared to another PA/GF sample manufactured using the conventional Extrusion Compression Molding (ECM) process. The 0°/0° orientation (high alignment) samples show a 35% increase in stiffness and a 30% increase in strength compared to ECM samples. The 0°/90° samples demonstrate tailored performance through controlled fiber orientation across the thickness. Additionally, other orientation options, such as + 45°/−45° to accommodate torsional loading, can be achieved using this method, highlighting its versatility in tailoring mechanical properties.

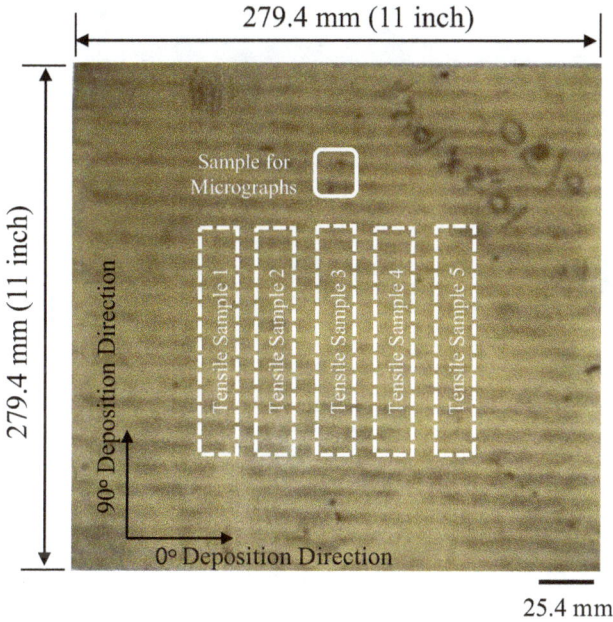

Figure 7.6: Consolidated AM preform after compression, showing deposition direction and panel dimensions (Courtesy of Oak Ridge National Laboratory, USA).

The AM-CM process enables the use of multimaterial preforms, as illustrated in Figure 7.9, offering enhanced functionality and performance [36]. Figure 7.9 demonstrates the capability to deposit carbon fiber-reinforced Acrylonitrile Butadiene Styrene (ABS) material (black, in Figure 7.9) in a complex shape, while the rest of the sample is printed using glass fiber-reinforced ABS (orange, in Figure 7.9). This highlights the versatility of the AM-CM process in achieving intricate designs and material distribution. Additionally, this method allows for the strategic placement of highly functional materials in specific locations, enhancing the overall performance of the parts. The AM-CM process supports parts fabrication using recycled and nonrecycled fibers, promoting sustainability. The ability to incorporate recycled materials not only reduces waste but also contributes to the development of environmentally friendly manufacturing practices. This method opens new possibilities for creating composite parts with tailored properties by combining different materials to achieve desired mechanical and functional characteristics, making the integration of multimaterial preforms and the strategic placement of high-performance materials a powerful tool for advancing composite manufacturing.

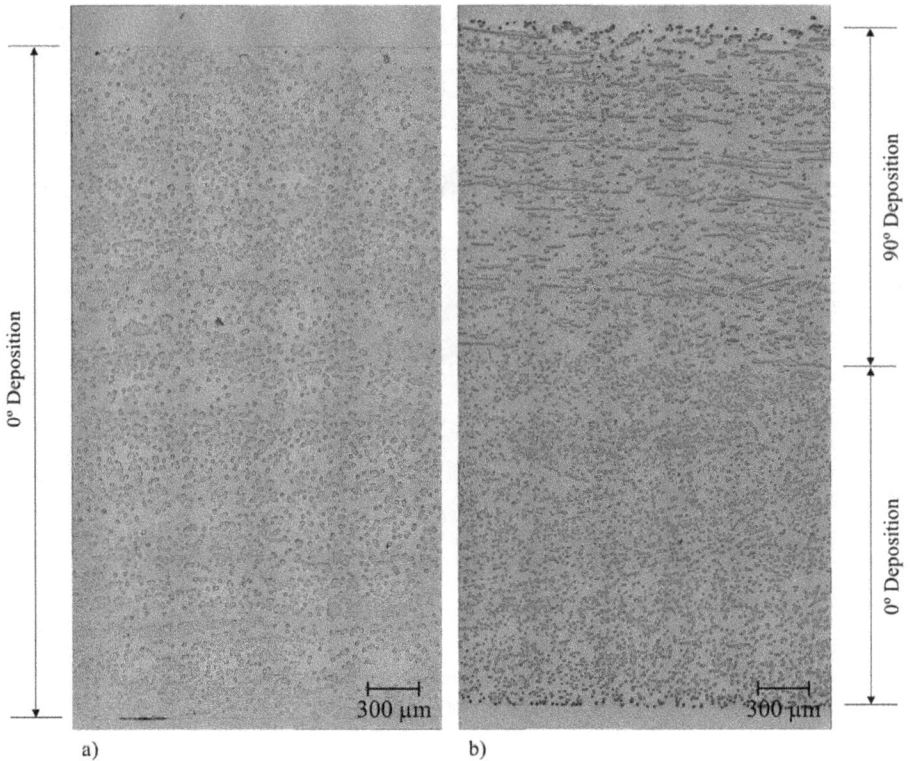

Figure 7.7: Microstructure of the cross section of consolidated panels manufactured by AM-CM; (a) In the 0°/0° sample, all fibers are aligned in the deposition direction, evident from their circular cross sections throughout the entire thickness, and (b) In the 0°/90° sample, one layer displays fibers with circular cross sections, while the 90° layer shows fibers with elongated ellipsoids and rectangular shapes (Courtesy: Oak Ridge National Laboratory, USA).

7.2.3 System development and implementation

The development of the AM-CM system involves integrating advanced hardware and software components to enable seamless operation and precise control. The system has three major hardware components: a fast-acting compression molding press, a robotic arm, and a high-throughput extruder (see Figure 7.10).

Robotic Arm: The system utilizes a 300 kg KUKA robot capable of fabricating intricate structures with high precision. A robot for large-scale additive manufacturing must prioritize precision, speed, and a large workspace. It needs stability to prevent vibrations, adaptability to different materials, robust control systems, and safety features. Easy maintenance and integration with Computer-Aided Design (CAD) and Computer-

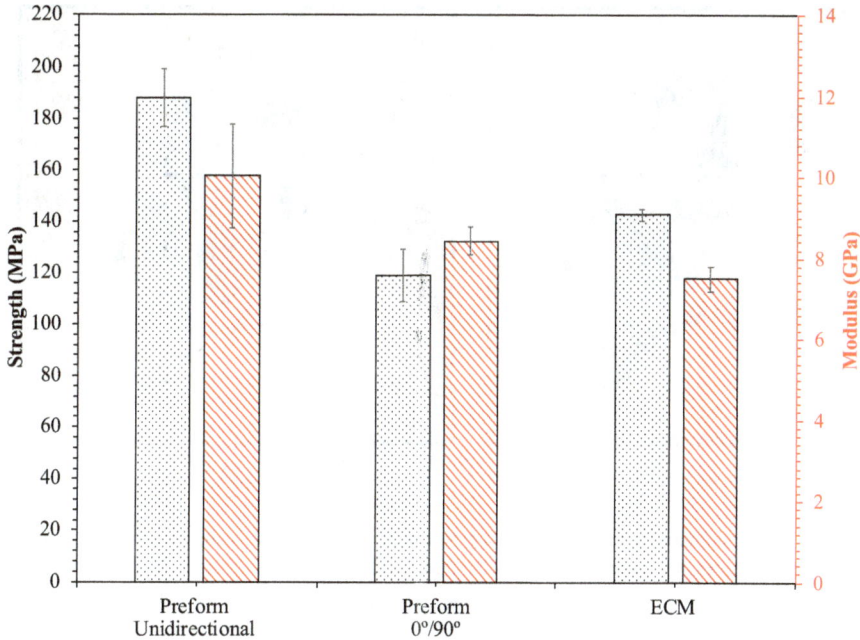

Figure 7.8: Mechanical properties (strength and stiffness) of AM-CM samples compared to those manufactured using the conventional ECM process. The data demonstrate the capability to control mechanical properties through precise management of the deposition process (Courtesy of Oak Ridge National Laboratory, USA).

Aided Manufacturing (CAM) systems also ensure efficiency and cost-effectiveness. The robot is equipped with a versatile tool changer capable of seamlessly switching between different end effectors. This allows for a wide range of fabrication tasks, such as a precision tape layup head for overmolding, a secondary material deposition extruder, and pick and place operations. This capability is crucial for the AM-CM process, as it allows for integrating multiple manufacturing techniques in a single, streamlined operation, significantly improving the overall productivity and quality of the composite parts produced.

AM Extruder: The robot is fitted with an extruder, capable of a deposition rate of 68 kg/h (150 lb/h). The extruder incorporates a general-purpose screw but can be equipped with different screw types to process long fibers, high-temperature materials, or materials requiring thorough mixing and dispersion. The extruder can be fitted with a variable-diameter nozzle, ranging from 2.54 mm to 12.7 mm (0.1 inch to 0.5 inch), and variable lengths, ranging from 12.7 mm to 150 mm (0.5 inch to 6 inches). The shape, size, and length of the nozzle significantly influence fiber alignment during the deposition process. The material, supplied in pellet form, is fed into the extruder through a delivery system connected to a dryer, which pre-dries the material

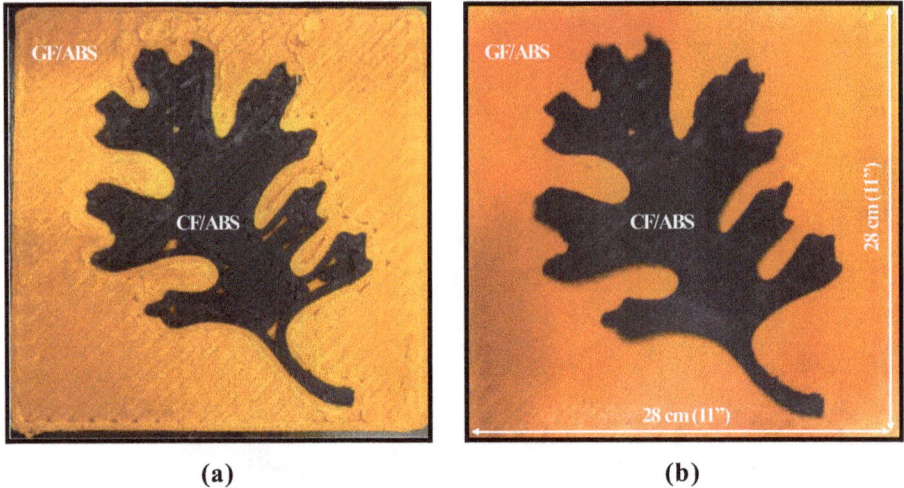

Figure 7.9: AM-CM multimaterial preforms; (a) AM preform before compression and consolidation, and (b) preform after compression molding and consolidation. This demonstrates the capability to deposit multimaterials, with carbon fiber-reinforced ABS (black) forming complex shapes, while the remainder of the sample is printed using glass fiber-reinforced ABS (orange) [Courtesy: Oak Ridge National Laboratory, USA].

to ensure optimal processing conditions. This setup allows for precise control over the deposition process, ensuring high-quality, and well-aligned composite parts.

Fast-Acting Compression Molding Press: The AM system is integrated with a fast-acting 500-ton hydraulic compression molding press (Trinks 500 Ton Down-Acting Compression Press with shuttle table), capable of achieving a closing speed of 12,200 mm/min (480 inches/min), a clamping speed selectable between 127 and 1,168 mm/min (5 to 46 inches/min), and accommodating molds up to 1.2 m × 1.2 m (48 inches × 48 inches) in size. The press features a shuttle table that can move the lower side of the mold out of the press, allowing easy access for the AM system to deposit material on top of the mold. If the shuttle option is not feasible, careful consideration should be given to the daylight opening of the press and the kinematics of the robot to ensure it can print inside the press without any interference or collision points. The shuttle uses variable hydraulic pressure (maximum of 1500 PSI) to control the shuttle velocity into the press. The fast-acting functionality of the press is crucial, as pressure needs to be applied to the preform before the material begins to solidify on the mold. The molds and the press platens are equipped with temperature controls to maintain optimal processing conditions.

The system has several safety features, making safety considerations paramount in its design and operation to ensure operator protection and system reliability. In the event of emergencies or unforeseen hazards, the system incorporates easily accessible emer-

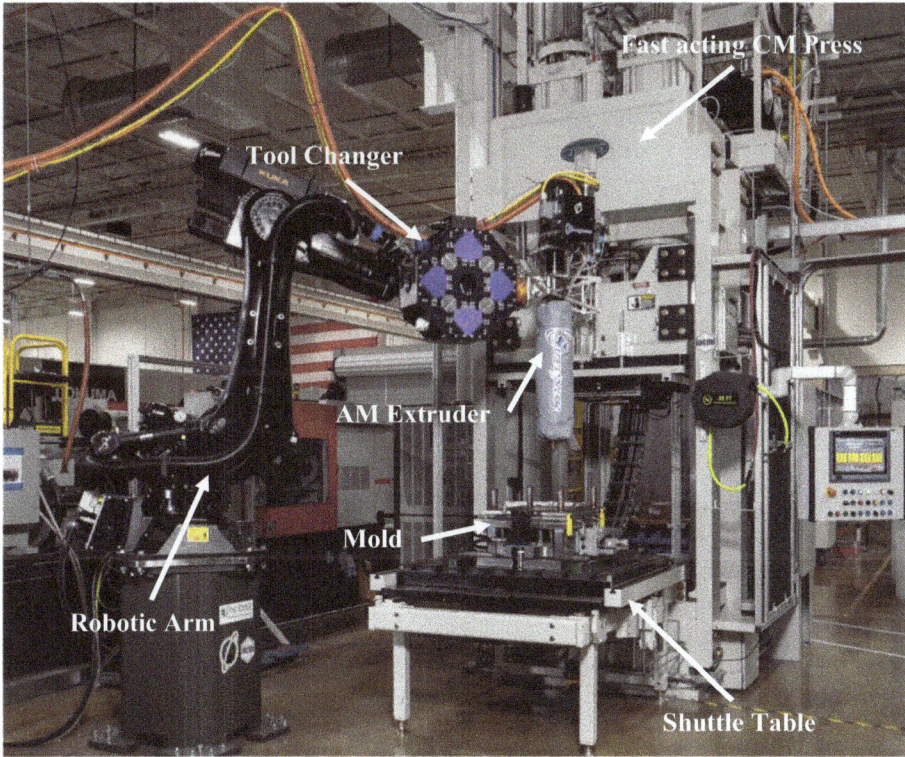

Figure 7.10: Developed integrated AM-CM system at ORNL's Manufacturing Demonstration Facility (Courtesy of Oak Ridge National Laboratory, USA).

gency stop mechanisms, providing immediate cessation of operations to mitigate risks effectively. Real-time monitoring capabilities continuously track operational parameters, enabling early detection of anomalies and triggering alerts for prompt intervention. Furthermore, safety interlocks enforce conditions under which specific actions can safely proceed, preventing unsafe operations. Physical barriers and shields further enhance safety by providing robust protection for operators and bystanders from moving parts and potential material ejections during the additive manufacturing and compression molding processes. As advancements continue, additional safety measures are continuously integrated to uphold stringent safety standards, ensuring the system operates reliably and securely, while optimizing productivity and operational efficiency.

The AM-CM system also integrates a software interface that optimizes toolpath planning and process control for additive manufacturing and compression molding. This software utilizes advanced algorithms to generate optimized toolpaths, tailored to part geometry and material properties, minimizing production time and material waste, while ensuring high precision. It enables comprehensive process simulation and optimization that provide opportunities to refine parameters such as deposition speed for optimal part quality before production starts. Real-time monitoring of critical parameters like temperature and pressure ensures consistent part quality and prevents defects during manufacturing.

The process begins with a detailed investigation of the part of interest using Finite Element Analysis (FEA) to determine the loading conditions and define the necessary microstructural architecture to withstand these loads. This analysis guides the slicing of the part with a tooling path corresponding to the precise material deposition on the mold. In this system, the part and mold are loaded into the software, which generates the tool path and deposition strategy. The sliced part is then transferred to robotic software that recommends and commands the robotic motion. The AM system prints the material in the desired configuration on the mold, which can involve one, two, or three layers, depending on the part's thickness and the required mechanical properties through the thickness. The deposition path of each layer is precisely controlled to ensure optimal fiber alignment. After the preforms are printed, the mold is shuttled inside the CM press. The press applies pressure and holds it for a period ranging from 1 min to 5 min, depending on the material type and the heating profile. This pressure application ensures the material consolidates effectively, resulting in a fully formed composite part. Once consolidation is complete, the part is ejected from the mold and is ready for use. This method opens new possibilities for creating composite parts with tailored properties, combining different materials to achieve desired mechanical and functional characteristics.

7.3 Molded parts performance evaluation

The integrated system was utilized to manufacture parts for various applications. The performance of these composite parts was assessed through mechanical testing and microstructural analysis. Key properties such as tensile strength, flexural modulus, impact resistance, and other mechanical characteristics were measured to evaluate the effectiveness of fiber alignment and material consolidation. X-ray microcomputed tomography (XCT) was utilized to analyze fiber orientation distribution and volumetric porosity, offering insights into microstructural integrity. Detailed information on the experimental setup and findings from a study comparing the performance of the AM-CM system with additive manufacturing and ECM techniques can be found in [36].

7.3.1 Mechanical properties and microstructural analysis

Figure 7.11 presents the XCT scan data, illustrating the fiber orientation distribution across AM, ECM, and AM-CM techniques. Additionally, it shows statistical distributions of the percentage of fibers at various orientation angles, divided into nine subsections with 10° increments, ranging from 0° to 90° [36]. The results indicated that the AM-CM system achieved high fiber alignment, similar to the AM process, with approximately 82% of fibers oriented within the 0° to 20° range in the θ direction [36]. In contrast, ECM samples exhibited minimal preferred orientation, with only about 7% of fibers within the 0° to 20° range in the θ direction. The high fiber orientation observed in AM samples is attributed to the extrusion deposition process [46], which aligns fibers along the deposition direction due to elongation forces during shear flow. However, ECM lacks control over fiber orientation across part thickness, leading to more random fiber orientation.

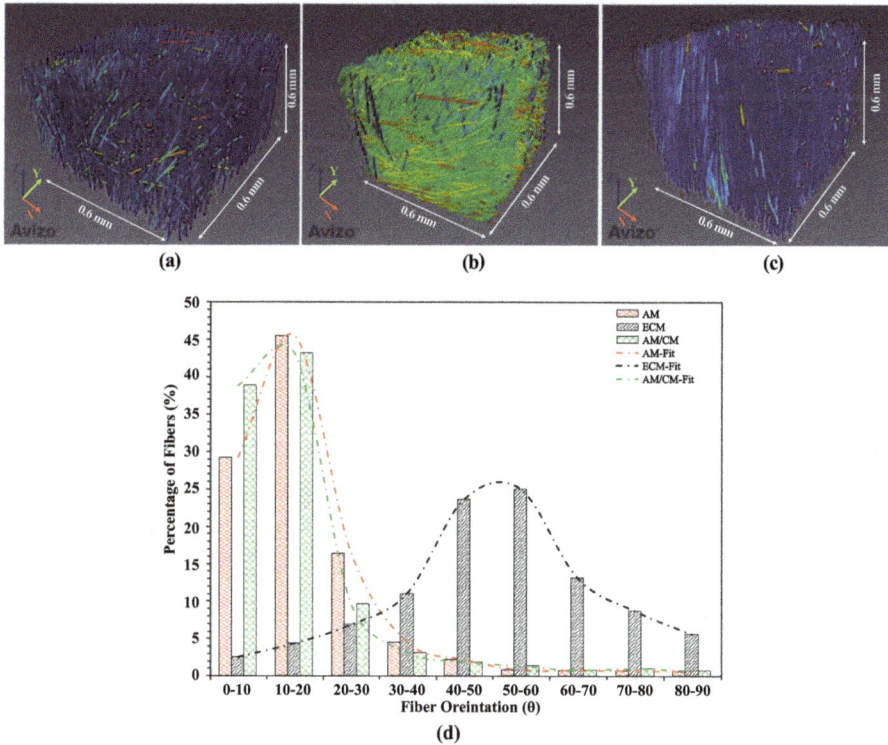

Figure 7.11: X-ray CT scan data showing FOD for all manufacturing processes: (a) AM, (b) ECM, and (c) AM-CM; and (d) Statistical distribution for the percentage of fiber in different orientation angles. Total fibers counted between 0° and 90° were divided into nine subsections of 10° increments [36].

Porosity analysis of the CF/ABS coupons from AM, ECM, and AM-CM samples also revealed distinct differences (Figure 7.12). AM samples showed an average volumetric porosity of 3.79% within the bead, indicating higher overall porosity due to unmeasured macroporosity between layers. ECM and AM-CM samples exhibited smaller porosity sizes, less than ~ 7 µm (the carbon fiber diameter). AM-CM samples reduced void content from 3.79% in preforms to 1.91% after compression molding [36]. Nota-

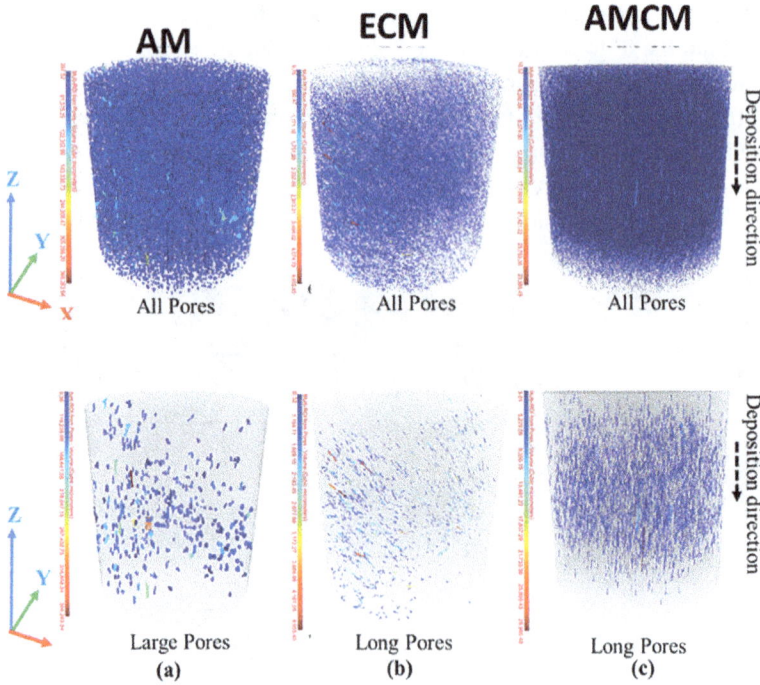

Figure 7.12: XCT scans showing all and large pores in samples from different manufacturing processes: (a) AM, (b) ECM, and (c) AM-CM [36].

bly, long pores aligned along fiber directions were observed in ECM and AM-CM samples, which were attributed to pore elongation during material flow [47, 48].

The mechanical properties of the AM-CM produced samples were consistent with the microstructural analysis findings. Neat samples, without fibers, exhibited similar modulus across all manufacturing methods. However, AM-CM and ECM neat samples showed higher tensile strength than AM, with ECM demonstrating a 27.67% increase due to the reduced void content from the compression molding process. For fiber-reinforced samples, AM showed a 28.89% higher modulus than ECM due to superior fiber alignment, and AM-CM exhibited a 35.27% improvement over ECM and 44.65% over AM. Additionally, fiber-reinforced samples from ECM and AM-CM demonstrated higher tensile strength than those manufactured using the AM process, with AM-CM showing a 11.15% improvement over ECM [36]. This behavior was also observed in flexural properties, with AM samples exhibiting better flexural performance than ECM. These properties were further enhanced by compression molding in AM-CM, resulting in a 28.6% increase in flexural strength and a 74.3% increase in flexural modulus. For impact properties, AM-CM and AM samples showed improvements of 46% and 5% over ECM, respectively. Overall, AM-CM samples demonstrated superior mechanical properties due to reduced void content and better fiber alignment, making it a promising method for manufacturing enhanced composite materials.

7.4 Applications for the AM-CM system

The AM-CM system offers significant advantages in industries such as automotive, aerospace, and consumer goods. By precisely controlling fiber orientation and microstructure, it produces lightweight, high-strength components with tailored properties. This system provides innovative solutions for composite manufacturing, especially in automotive and transportation applications where rapid production of high-performance parts is essential. With a cycle time as short as 5 min per part that are up to 0.3 m (1 ft.) × 0.3 m (1 ft.), the AM-CM system meets the high throughput demands necessary for efficient production. The AM-CM can facilitate selective reinforcement, representing a significant advancement in composite manufacturing, offering tailored solutions to enhance the performance and efficiency of complex parts. By embedding high-strength fibers or functional additives in specific regions, manufacturers can optimize material use and enhance localized mechanical properties such as strength, stiffness, and durability. This approach enables the creation of lightweight structures with reinforced load-bearing areas, reducing overall weight, while maintaining structural integrity. Additionally, AM-CM can integrate functional features, such as conductive pathways, thermal management structures, or customized surface properties, directly during the manufacturing process. This capability streamlines production, facilitates complex geometries, and optimizes material distribution based on performance needs.

The precise digital control in the AM-CM process allows for its effective use in joining dissimilar hybrid materials. Controlling the deposition and fiber architecture at the joining area is crucial for optimizing load-bearing performance. A study by Pokkalla et al. [49] demonstrated the potential of the AM-CM process to enhance the mechanical properties of polymer-metal hybrid composite materials. In this study, a maraging steel lattice was fabricated using AM, followed by compression molding with carbon fiber-reinforced polyamide-6,6 (40% wt CF/PA66) preforms. The resulting hybrid composites exhibited tensile strengths 2–3 times higher than those achieved by other bonding methods. Fracture analysis indicated that failure occurred primarily at stress concentration points, underscoring the effectiveness of this approach in creating mechanically interlocked hybrid structures, without the need for adhesives or complex surface treatments. Moreover, the process can be coupled with continuous carbon fiber overmolding to enhance composite joints. These fibers can be integrated either atop the mold or preprinted and overlaid during compression molding, resulting in built-in stiffeners along the load direction, and achieving unprecedented bend radii. This enhancement can improve the flexural strength by 57% and flexural modulus by 49%, while reinforcing joints between parts through overmolding [50].

AM-CM can be utlized for producing complex out-of-the-plane parts due to the multi-axis capabilities of robotic systems, which offer flexibility compared to gantry-based systems. Direct printing onto intricate molds allows precise control of fiber orientation, whether highly aligned or randomly oriented, based on structural and warpage tolerance requirements. This method also enables optimization of overfilling or underfilling mold areas, effectively reaching deep draws and complex shapes that are otherwise inaccessible. Additionally, it allows for variable thickness control within a single part. As demonstrated, rotor blades for drones and sections of concrete chutes were successfully fabricated for Orbital Composites and Oshkosh Truck, respectively, using glass-filled polycarbonate composites (see Figure 7.13). With industries increasingly demanding lighter, stronger, and more versatile components, AM-CM provides an advanced solution, driving innovation across aerospace, automotive, biomedical, and other high-performance sectors.

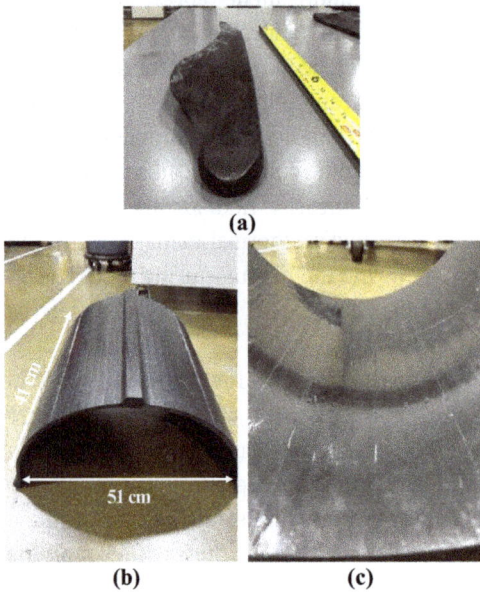

Figure 7.13: Out-of-the-plane complex AM-CM parts; (a) Rotor blade for a functional drone for Orbital Composites, and (b) & (c) Subsection of a concrete chute for Oshkosh Trucks using glass-filled polycarbonate composite (Courtesy: Oak Ridge National Laboratory, USA).

7.5 Future directions

Future advancements in AM-CM technology are focused on enhancing process efficiency, expanding material versatility, and improving part performance. Key areas of development include creating a digital twin of the system for precise process control, exploring novel materials such as long fiber thermoplastics, and advancing process monitoring and automation technologies. Integrating a digital twin framework, coupled with Artificial Intelligence (AI) and Machine Learning (ML) algorithms, will enable predictive modeling and optimization of the AM-CM process. This integration will enable closed-loop system integration for real-time adjustments and quality control, facilitating smart manufacturing practices that ensure consistent and optimized production outcomes.

A current limitation of AM-CM is the maximum size of parts it can produce, primarily due to thermal challenges that can cause early solidification of the deposited material before the compression molding process, as well as mold size constraints. To address the thermal challenge, implementing an auxiliary heating system is essential for better retaining the deposited part's heat. To overcome mold size constraints, technologies are being developed for creating larger parts through sequential additive compression overmolding. This approach, analogous to the continuous compression molding process but tailored for discontinuous reinforced composites, involves joining parts in a sequential manner.

7.6 Conclusion

The importance of integrated manufacturing processes lies in their ability to streamline manufacturing workflows, reduce lead times, and enhance material utilization, thereby driving efficiency and cost-effectiveness. Integrating additive manufacturing and compression molding in the AM-CM system represents advancement in composite digital manufacturing technology. AM offers benefits such as reduced material waste, the capability to produce advanced geometries, and the ability to deposit fibers that are uniformly dispersed and optimally oriented, leading to significantly reduced lead times. Conventional composites manufacturing processes like compression molding contribute to fast cycle times, removing voids, and high-quality surface finishes of parts. This combination is ideal for low-cost, high-volume applications. This synergy between the manufacturing processes not only enhances the mechanical properties of the produced parts but also optimizes the entire manufacturing process. This integrated approach allows for the creation of complex, high-performance composite structures that were previously unattainable. The potential to sequentially overmold larger parts addresses current limitations in part size and opens new possibilities for large-scale applications.

Acknowledgment: The authors acknowledge the support from the US Department of Energy (DOE), Office of Energy Efficiency and Renewable Energy, and Advanced Materials and Manufacturing Office. This manuscript has been authored by UT-Battelle, LLC, under contract DE-AC05-00OR22725 with the US Department of Energy (DOE). The US government retains and the publisher, by accepting the article for publication, acknowledges that the US government retains a nonexclusive, paid-up, irrevocable, worldwide license to publish or reproduce the published form of this manuscript, or allow others to do so, for US government purposes. DOE will provide public access to these results of federally sponsored research in accordance with the DOE Public Access Plan (http://energy.gov/downloads/doe-public-access-plan).

References

[1]   N.T. Tuli, S. Khatun, A.B. Rashid Unlocking the future of precision manufacturing: A comprehensive exploration of 3D printing with fiber-reinforced composites in aerospace, automotive, medical, and consumer industries, Heliyon, 2024, 10(5): e27328.

[2]   A.A. Firoozi, A.A. Firoozi, *A systematic review of the role of 4D printing in sustainable civil engineering solutions.* Heliyon, 2023. **9**(10): p. e20982.

[3]   S. Alkunte et al. Advancements and challenges in additively manufactured functionally graded materials: A comprehensive review, Journal of Manufacturing and Materials Processing, 2024, 8: DOI: 10.3390/jmmp8010023.

[4]   B.W. Grimsley et al. High-rate aircraft manufacturing: In-situ consolidation automated fiber placement of thermoplastic composites for high-rate aircraft manufacturing. SAMPE Journal, 2022, 58(4): 38–54.

[5]   J.-M. Lee, B.-M. Kim, D.-C. Ko Development of vacuum-assisted prepreg compression molding for production of automotive roof panels, Composite Structures, 2019, 213: 144–152.

[6]   G.L. Hahn, T.K. Tsotsis Rapid High-Performance Molding (RAPM) for Small Parts, SAMPE 2019-Charlotte, NC, May 2019, 2019.

[7]   A. Aravand Composite injection overmoulding, in: Design and Manufacture of Structural Composites, Elsevier, 2023, pp. 323–345.

[8]   T. Link et al. Hybrid composites for automotive applications-development and manufacture of a system-integrated lightweight floor structure in multi-material design, in: Proceedings of the SPE 19th Annual Automotive Composites Conference & Exhibition, ACCE, 2019.

[9]   R. Brooks Forming technology for thermoplastic composites, Composites Forming Technologies, 2014, 2007: 256–276.

[10] T. Kulhan et al. Fabrication methods of glass fibre composites – A review, Functional Composites and Structures, 2022, 4(2): 022001.

[11] R. Thompson Manufacturing Processes for Design Professionals, Thames & Hudson, New York, NY, 2007.

[12] D.K. Rajak et al. Fiber-reinforced polymer composites: Manufacturing, properties, and applications, Polymers, 2019, 11(10): 1667.

[13] A. Diniță et al. Advancements in fiber-reinforced polymer composites: A comprehensive analysis, Polymers, 2023, 16(1): 2.

[14] M. Valente, I. Rossitti, M. Sambucci Different production processes for thermoplastic composite materials: Sustainability versus mechanical properties and processes parameter, Polymers (Basel), 2023, 15(1): 242.

[15] B. Gangil et al. Introduction to thermosetting polymer composites: applications, advantages, and drawbacks, in: Dynamic Mechanical and Creep-Recovery Behavior of Polymer-Based Composites, Elsevier, 2024, pp. 11–19.

[16] M. Biron Thermosets and Composites: Material Selection, Applications, Manufacturing and Cost Analysis, Elsevier, Amsterdam, 2013.

[17] R.R. Nagavally Composite materials-history, types, fabrication techniques, advantages, and applications, Int. J. Mech. Prod. Eng, 2017, 5(9): 82–87.

[18] M. Barile et al. Thermoplastic composites for aerospace applications, in: Revolutionizing Aircraft Materials and Processes, Springer, 2020, pp. 87–114.

[19] M. Biron Thermoplastics and Thermoplastic Composites, William Andrew, 2018.

[20] S. Hummel et al. Analysis of mechanical properties related to fiber length of closed-loop-recycled offcuts of a thermoplastic fiber composites (organo sheets), Materials, 2022, 15(11): 3872.

[21] B. Richter, B. Neitzel, F. Puch Extrusion as an energy-efficient manufacturing process for thermoplastic organosheets, in: Journal of Materials Research Proceedings, Materials Research Forum LLC, Germany, 2023.

[22] D. Kaige et al. Preparation and properties of cross- linked thermoplastic polyethylene (TPEXa) pipes, China Plastics, 2023, 37(1): 8.

[23] A.A. Abreu, S.I. Talabi, A. De Almeida Lucas Influence of nucleating agents on morphology and properties of injection-molded polypropylene, Polymers for Advanced Technologies, 2021, 32(5): 2197–2206.

[24] P. Mallick Thermoplastics and Thermoplastic–Matrix Composites For Lightweight Automotive Structures, in: Materials, Design and Manufacturing for Lightweight Vehicles, Elsevier, 2021, pp. 187–228.

[25] M. Dannemann et al. Damping behavior of thermoplastic organic sheets with continuous natural fiber-reinforcement, Vibration, 2021, 4(2): 529–536.

[26] A. Żur et al. Preliminary study on mechanical aspects of 3D-printed PLA-TPU composites, Materials, 2022, 15(7): 2364.

[27] J. Halpin Stiffness and expansion estimates for oriented short fiber composites, Journal of Composite Materials, 1969, 3(4): 732–734.

[28] K. Olonisakin et al. Key improvements in interfacial adhesion and dispersion of fibers/fillers in polymer matrix composites; focus on pla matrix composites, Composite Interfaces, 2022, 29(10): 1071–1120.

[29] S. Khandelwal, K. Y. Rhee Recent advances in basalt-fiber-reinforced composites: Tailoring the fiber-matrix interface, Composites Part B: Engineering, 2020, 192: 108011.

[30] A.A. Jaber et al. Effect of fiber sizing levels on the mechanical properties of carbon fiber-reinforced thermoset composites, Polymers, 2023, 15: DOI: 10.3390/polym15244678.

[31] R. Krishna, M. Manjaiah, C. Mohan Developments in additive manufacturing, in: Additive Manufacturing, Elsevier, 2021, pp. 37–62.

[32] V. Kishore, A.A. Hassen Chapter 6 – Polymer and Composites Additive Manufacturing: Material Extrusion Processes, in: J. Pou, A. Riveiro, J.P. Davim, (Eds.), Additive Manufacturing, Elsevier, Amsterdam, 2021, pp. 183–216.

[33] Y. Huang, M.C. Leu Frontiers of Additive Manufacturing Research and Education, William Andrew. Cambridge, 2014.

[34] Y. Huang et al. Additive manufacturing: Current state, future potential, gaps and needs, and recommendations, Journal of Manufacturing Science and Engineering, 2015, 137(1): 014001.

[35] S.L. Ford Additive manufacturing technology: Potential implications for US manufacturing competitiveness, Journal of International Economic, 2014, 6: 40.

[36] V. Kumar et al. High-performance molded composites using additively manufactured preforms with controlled fiber and pore morphology, Additive Manufacturing, 2021, 37: 101733.

[37] B.G. Compton et al. Thermal analysis of additive manufacturing of large-scale thermoplastic polymer composites, Additive Manufacturing, 2017, 17: 77–86.

[38] T. Ward; Available from: https://umaine.edu/news/blog/2024/04/23/umaines-new-3d-printer-smashes-former-guinness-world-record-to-advance-the-next-generation-of-advanced-manufacturing/.

[39] N. Van de Werken et al. Additively manufactured carbon fiber-reinforced composites: State of the art and perspective, Additive Manufacturing, 2020, 31: 100962.

[40] T. Smith et al. Large scale additive manufacturing: Dual material system for polymer large scale additive manufacturing, SAMPE Journal, 2021, 57(6): 1–10.

[41] A.A. Hassen Non Destructive Evaluation and Characterization of Continuous and Long Fiber Reinforced Thermoplastic Composites, The University of Alabama at Birmingham, USA, 2015.

[42] W. Harnnarongchai, S. Patcharaphun Prediction of weldline strength for injection molded short-glass-fiber composites, Materials Today: Proceedings, 2023, 77: 1122–1126.

[43] S. Li et al. Additive manufacturing-driven design optimization: Building direction and structural topology, Additive Manufacturing, 2020, 36: 101406.

[44] W. Ogierman, G. Kokot A study on fiber orientation influence on the mechanical response of a short fiber composite structure, Acta Mechanica, 2016, 227(1): 173–183.

[45] S. Yuan et al. Additive manufacturing of polymeric composites from material processing to structural design, Composites Part B: Engineering, 2021, 219: 108903.

[46] H.L. Tekinalp et al. Highly oriented carbon fiber–polymer composites via additive manufacturing, Composites Science and Technology, 2014, 105: 144–150.

[47] D. Yang et al. Fibre flow and void formation in 3D printing of short-fibre reinforced thermoplastic composites: An experimental benchmark exercise, Additive Manufacturing, 2021, 37: 101686.

[48] A. Vaxman et al. Void formation in short-fiber thermoplastic composites, Polymer Composites, 1989, 10(6): 449–453.

[49] D.K. Pokkalla et al. A novel additive manufacturing compression overmolding process for hybrid metal polymer composite structures, Additive Manufacturing Letters, 2023, 5: 100128.

[50] J.B. Tyler Smith, R. Walker, V. Kumar, D. Nuttall, R. Ogle, J. Charron, C. Duty, V. Kunc, A.A. Hassen Continuous fiber 3D printing for compression overmolding, in: Camx 2023, SAMPE: ATLANTA, GA, 2023.

Index

https://doi.org/10.1515/9783111019543-008

www.ingramcontent.com/pod-product-compliance
Lightning Source LLC
Chambersburg PA
CBHW081539220326

41598CB00036B/6494